Traces of (Un-) Sustainability

This book is part of the Peter Lang Political Science, Economics, and Law list.
Every volume is peer reviewed and meets
the highest quality standards for content and production.

PETER LANG
New York • Bern • Berlin
Brussels • Vienna • Oxford • Warsaw

Peter Graham

Traces of (Un-) Sustainability

Towards a Materially Engaged Ecology of Mind

PETER LANG
New York • Bern • Berlin
Brussels • Vienna • Oxford • Warsaw

Library of Congress Cataloging-in-Publication Data

Names: Graham, Peter (Sociologist) author.
Title: Traces of (un-) sustainability: towards a materially engaged ecology of mind / Peter Graham.
Description: 1 Edition. | New York: Peter Lang, 2020.
Includes bibliographical references and index.
Identifiers: LCCN 2019055024 | ISBN 978-1-4331-7666-1 (hardback: alk. paper)
ISBN 978-1-4331-7672-2 (ebook pdf)
ISBN 978-1-4331-7673-9 (epub) | ISBN 978-1-4331-7674-6 (mobi)
Subjects: LCSH: Sustainability—Social aspects. | Economic development—Moral and ethical aspects. | Human ecology.
Classification: LCC HC79.E5 G68533 2020 | DDC 304.2—dc23
LC record available at https://lccn.loc.gov/2019055024
DOI 10.3726/b17003

Bibliographic information published by **Die Deutsche Nationalbibliothek**.
Die Deutsche Nationalbibliothek lists this publication in the "Deutsche Nationalbibliografie"; detailed bibliographic data are available on the Internet at http://dnb.d-nb.de/.

The paper in this book meets the guidelines for permanence and durability of the Committee on Production Guidelines for Book Longevity of the Council of Library Resources.

Printed in the United States of America

For my late mother, Martha.

In the last few days, people have asked me, "What do you mean, ecology of mind?" Approximately, what I mean is the various kinds of stuff that goes on in one's head and in one's behavior—in dealing with other people, walking up and down mountains, getting sick, and getting well. All that stuff interlocks and, in fact, constitutes a network which, in the local language, is called mandala. I'm more comfortable with the word "ecology," but they're very closely overlapping ideas. At the root, it is the notion that ideas are interdependent, interacting, that ideas live and die. The ideas that die do so because they don't fit with the others. You've got a sort of complicated, living, struggling, cooperating tangle like what you'll find on any mountainside with the trees, various plants, and animals that live there—in fact, an ecology. Within that ecology, there are all sorts of main themes that one can dissect out and think about separately. There is always, of course, violence to the whole system if you think about the parts separately; but we're going to do that if we want to think at all, because it's too difficult to think about everything at once. So I thought I would try to unravel for you some of the ecology, something of the position and nature of the sacred in the ecological system (Bateson, 1991, p. 265).

Figure P.1. Bee Balm. Photo courtesy of Gary Kennedy, 2017.

TABLE OF CONTENTS

ILLUSTRATIONS

PREFACE

The theoretical framework of a materially engaged ecology of mind, as set out in this book, is quite obviously an amalgamation and remixing of many others' ideas. I have had the very great good fortune to have been able to sample such ideas widely, taking the choicest bits from here and there, integrating, filtering, and synthesizing as I went along. The contributions of some, such as the writings of Gregory Bateson or Lambros Malafouris or the generous conversations with Mick Smith or Rena Upitis, for example, were massive and irreplaceably constitutive of the finished work. Yet, who could say with any degree of certainty that even the turkeys I always looked for, and usually found, from the bus window during my weekly commutes between Pointe Claire and Kingston, were not, in some sense at least, the prerequisite inspiration that caused the final pieces of the puzzle to fall into place for me? Where I grew up, wild turkeys made their incredible comeback during my teenage years. I was walking with Toby, our Brittany spaniel, when I first encountered a wild turkey sometime in the early 1970s. The experience was magical and indelibly set in my psyche ever after. My affection for turkeys in the landscape continues to inspire me, so does an entire world of ideas and material things, some dangerous, to be sure, and others so contagiously vivacious and inspiring that it is difficult to imagine even the possibility of life without them.

In truth, my own creative contribution has been relatively minor, akin to that of a sort of conduit, or a kitchen appliance perhaps with a sort of emulsifying attachment. The world, in some sense, passed through me to call forth this book. The "me" often simply was lost in the process and, of course, as I now realize with the benefit of hindsight, that was exactly the great benefit of the whole exercise. My wish for readers is that you, too, will begin to get lost in a lively and enchanted material world; a world of books, grasses, sparrows, stomachs, eyeballs, trees, teachers, chironomids, kitchen sinks, and so much more. Our own health and liveliness depends primarily upon how skillfully we come to entangle, extend, and distribute ourselves within our lively and beautiful world.

The relationship we have with the world is always reciprocal. Take any book, like any tree or other living thing: the world oozes back out in discernible patterns of words, growth rings, habits, tendencies, or rituals. Sustainability means then that the world seeping back to us through our artifacts, through our designing designs, should be a lively and sustainable world and not simply a store of resources or a potential site for a toxic waste dump, not a paint-by-number project in a story of recovery and rehabilitation after some mythic event resulted in a broken world. Broken-world stories can only break us, as a people unwittingly transmitting brokenness reflexively back upon ourselves. Bateson made the point that "we become what we pretend," but I would respectfully amend that insightful observation to the following: We become what we ask the world to pretend us to be.

The world generally indulges us, at least to a point. We cannot have our cake and eat it too. If we cut down the trees, for example, the spirits may very well make a home in the roman columns designed for their habitation. Where else could they reside after all? But they will not still then be the same spirits. They will not comport themselves as before. They will not be bound to and intertwined with the same world. Their mood will change, and they may well become restless. The previous bond between peoples and tree spirits may be broken. We will then become the peoples we inadvertently asked the world to make us—peoples without a practiced spiritual connection to a wondrously unknowable world, a world with tree spirits … and trees.

I also humbly acknowledge and wish to express my gratitude for the important wisdom revealed to me in the many examples of Indigenous knowledges I have come across in my studies. I have tried not to misappropriate or misrepresent those knowledges in my following discussions. In this book, I do not seek to become an advocate for the adoption of one Indigenous knowledge or another. The examples of traditional knowledge I use in this book are intended as examples of individual aspects of complex knowledge systems compared to other individual aspects of other knowledge systems. The examples are in no way intended to reduce Indigenous knowledge to a simple iconic expression. My work also does not seek to put forward an alibi or excuse for the many cruel injustices inflicted in the past and continuing to be inflicted upon Indigenous peoples by settlers and settler cultural tools and meaning systems. I do not presume to have anything to teach Indigenous peoples. Yet, I do hope my work may be helpful to Indigenous peoples and that they might take some comfort in the slow learning on the part of settlers that this book is intended to represent. My goal is to provide healing of some sort to the Cartesian-infected social mind, to turn Western science upon itself, so that the toxicity of that historical people-meaning-system-environment configuration might be reduced. Although this book does not broach the topics, it might provide useful tools for addressing such difficult subjects as collective guilt and reconciliation. Just as the person our children will grow to become remains largely unknowable at birth, my aspirations for what this book will eventually become remain largely, at this point, beyond my control and its origins beyond my understanding.

And so, I want to begin at this starting point by recognizing and honoring where this book really comes from and its many authors, including the turkeys, trout, and strawberries that hopefully seep out from the pages that follow, exercising their own unique forms of agency. I am also especially indebted and grateful to Professor Mick Smith of the Queen's University School of Environmental Studies, who accompanied me on the whole journey from start to finish. Mick's empathetic and thoughtful guidance, his confidence in my ability to complete the work, and the many reflective, inspiring, and meaningful conversations we shared have meant the world to me. Special thanks also go to Rena Upitis, who mentored and guided me along the way. Graham Whitelaw also continued to inspire me with his endless generosity and wise counsel. Ingrid Stefanovic, Molly Wallace, and Ryan Danby all provided invaluable feedback on the initial draft, and I deeply appreciate the very thoughtful and constructive criticism they offered. I also want to acknowledge and thank Peter Lang's anonymous reviewer as well as Ann Dale and John Barry, who agreed to review the manuscript for the record.

I must also express my most sincere appreciation to the entire School of Environmental Studies at Queen's University. A better home base for my work would be simply unimaginable. Special thanks go to Julie Adams, Paul Bass,

Steven Brown, Karen Depew, Alice Hovorka, Myra Hird, Heather Jamieson, Colin Khan, Cassandra Kuyvenhoven, and Scott Lougheed.

There were simply far too many people who helped me get to the starting line, to mention them all, but people who had an exceptional influence on my formation as a scholar prior to my arrival at Queen's University include Adeela Arshad-Ayaz, Peter G. Brown, Ailie Cleghorn, Maguerite (Margie) Mendell, Ayaz Naseem, Katja Neves, David Newhouse, Daniel Salé, and Eric Shragge. Authors whose work made an especially deep impression on me include (but are in no way limited to) Gregory Bateson, Nicole Boivin, Kenneth Boulding, Jerome Bruner, Ursula Franklin, Chris Gosden, Jean Lave, Lambros Malafouris, Lev Vygotsky, and James Wertsch.

My students in the classes I teach at Concordia University inspire me, and I want to acknowledge their contribution as well. I am often reminded of my own childhood and my own children by listening to my students. They help and inspire me to keep learning about our shared world. They remind me constantly of the stakes involved. Their comments, their questions, and their tireless striving give me hope. My teaching always focuses on the value (both positive and negative) of the tools transmitted through the normal university experience. My teaching practice, and the academic freedom it requires, has been facilitated by the outstanding assistance of Perry Calce and Rebecca Tittler.

Teri-Ann McDonald deserves special credit for the mediating role she played between the novice writer and the standards of my publisher, Peter Lang. Which is not to diminish in any way the role played by Peter Lang. Michelle Smith, in particular, has been incredibly helpful and supportive throughout the entire publishing process.

Finally, to Shirley, John, and Martha—this book is especially for and about you—thank you.

ABBREVIATIONS

CBA	cost-benefit analysis
MET	Material Engagement Theory
PTSD	post-traumatic stress disorder
ROI	return on investment
TLT	transformative learning theory
ZPD	zone of proximal development

· 1 ·

SITUATING THE TRACES OF (UN-) SUSTAINABILITY

When things go wrong, it is generally important to understand how and why they do so. Our understanding of any given situation forms the basis for any plan to get ourselves back on the tracks. However, it is also important to understand that any such understanding is always incomplete. Not only is it often very difficult, if not impossible, to discern whether things that seem to be going badly are secretly creating the context for things to go much better in future, but our understandings are also always limited to a kind of partial map, or a trace of the actual situation, in the past, present, and future. Ignorance is a normal part of the human condition, regardless of how heroically we struggle to overcome the inadequacy of our maps and our more-or-less (in-) adequate knowledges.

Any given knowledge, like any metaphor, represents the trace we choose as our Ariadne's thread to guide us forward, but each choice necessarily means forgoing other threads that would guide us in different directions. Some maps are clearly better than others, and they are better because they improve our situation and allow us to get out of the immediate trouble that induced us to seek out a map in the first place. However, maps are not necessarily better because they are more accurate or objective, rational, or evidence based. They are better if they help us get to, and stay in, a better place—if, we might say, they can deliver real sustainability. This is not to say that reason, logic, and evidence should be abandoned in designing our explanatory theories or accepting/rejecting conventional wisdoms—quite the contrary. But it is to say that explanatory theories and conventional wisdoms, however "rational" they may seem, can often only lead us further into the labyrinth.

Take, for example, an image of human beings as rational economistic and calculative profit maximizers. That specific model certainly makes a very bad map, in part because it does not reflect the readily available evidence that humans can be extremely cooperative, even when a time and place-specific historical inheritance induces us to pretend otherwise. More importantly, the *homo economicus* map makes a bad map *because* the territory (in this case our behavior) always begins to reflect the map. The map is not just a neutral "idea" or description of an external reality; rather, its employment plays a creative role in constituting that reality. It can therefore create more walls, false passages, and dead-ends. It has not been uncommon, throughout the history of Western civilization, for the prescriptions designed to solve a problem to actually make that problem a bigger problem and a more intractable problem. To understand this process, whereby conceptual map and reality are inextricably entwined, and how ideas (and ideologies) emerge from, and feed into, patterns of recurrent material unsustainability, we need, as the passage by Bateson that prepared the ground for this present work to grow and take shape suggests, to think about the material ecology of ideas, the historically specific confluence of peoples, places, mediational means, events, and environments. This requires that we cease thinking of ideas and world as separate realms mapping onto each other and begin to shift our focus toward seeing the social world as an ecology of meaningful material engagement. This introductory chapter will begin to explain what that might entail; Chapter 2 will further develop this topic and introduce a specific theoretical framework, that of Material Engagement Theory (MET).

Environment as Parent

In trying to explain how to get to a better place and stay there—how to nurture sustainability—this book will often employ the metaphor of a parent-child relationship to understand environmental/human relations (whether that environment is considered material, biosemiotic, phenomenological, social, political, psychological, economic, cultural, etc.). This is not to say that environment intentionally parents us, only that our experience of environment is often akin to our experience of a parent. This proposal might seem strange, yet our relationship with environment, our models and theoretical frameworks, and our sociocultural *action* is always based in and informed by one metaphor or another (Ingold, 2000; Weiss, 1995). The value of and human affinity for using such metaphorical tools to enhance our understanding is well established and well explained by others (see, for example, Boivin, 2008; Lakoff & Johnson, 1980; Lund & Benediktson, 2010; Morgan, 1997). My claim is that the parent-child metaphor could provide better results with respect to sustainability than, say, the currently dominant *homo economicus* model of isolated individuals competing for natural resources. It is easily demonstrable that the available evidence is on the side of the parent-child metaphor. Parents and environments co-constitute us. They leave indelible traces upon us. Parenting involves active physical/material engagement, and it is through such active engagement, not only with parents but also with the world more generally that human development and identity formation occurs.

Like any metaphor, the usefulness of the environment-is-like-a-parent metaphor has limits. However, the parent-child dyadic metaphor applied to person-environment relations provides important insights even beyond those achieved with other current metaphors such as "Gaia"—environment as living being (Lovelock, 1989)—or human being in "conversation" with environment (Lund & Benediktson, 2010). Lovelock's Gaia metaphor conjures an image of a solitary God-like being, alone in the universe but colonized by an emergent variety of "parasitic" free loaders, (although admittedly, a few may be better behaved than others and even aspire to become Gaia doctors (Lovelock, 1991). Conversation, likewise, misleads in that it implies a shared language used by two or more separate and more-or-less equal individuals. But the nature of the parent-child relationship does not rely on, nor is it necessarily altered by, the absence of a shared language. First, it does not *always* matter for the development of the child whether the child verbally understands the parent's perspective or not. Also, other means of communication between parent and child always emerge, evolve, and carry on before, during, and after the relationship is mediated by language.

When we engage with environment, whether beautiful sunset, laptop computer, or greenhouse gas, we are changed, if only a little, by the experience. The materiality of that experience acts neither like a God nor like the transmission and reception of information that one might associate with the term "conversation." Rather, our cognitive and emotional processes extend into the material, and the material extends into our brain/body. Parents and children enact this type of inter-extending process nearly continuously in the early stages of the child's life.

Humans are certainly not unique in having a parent-child type of relationship with environment either. Indeed, all life is world-parented and the forms and practices of parental guidance can be almost limitless, often exceeding the limits of the human capacity to comprehend. Light, heat, moisture, semiosis, community, and the diversity of materialities of environment all guide the development of all the world's various "children." Although the Neolithic advent of a sedentary and tool-mediated way of being in the world set the humans' relationship apart from that of other species in some ways, we would do well to not make too much of the distinction. Being different should certainly not be confused with being more evolutionarily fit for survival in some over-arching or teleological sense.

Consider the popular phrase of developmental psychology, that "there is no such thing as a baby: there is a baby and someone" (Winnicott, 1971, p. 70). The developmental psychology literature has been moving increasingly toward a perspective that maintains that the baby can only be properly understood as part of a baby-parent dyad (Bartolo, 1998; Clark, 2002; Fernandes Mendes, 2015; Kim & Kochanska, 2017; Renzi, Romberg, Bolger, & Newman, 2017). The baby develops *and* the parent develops always and only in a co-constitutive relationship with the other. What is also co-constituted in this process is a set of mediational means, and one important claim I want to begin this book with is that people, in general, can best be understood as person–cultural tool set–environment triads. Minds are always situated and mediated, meaning that human development always entails the two-directional mastery of a set of cultural tools (Wertsch, 1998). Babies, for example, begin to learn how to understand and express themselves using facial expressions. But while some facial expressions (e.g., smiling or crying) and some gestures such as pointing are universal, many others are culture-specific. I also want to make the point that the relationship is not one of equals. There is always a parent-specific role and a child-specific role. Environment is likewise not an equal partner in the person-environment relationship. Environment pre-exists and has an inherently greater capacity to exercise agency and to set the ground rules for the exercise of agency (Bateson & Bateson, 1987, pp. 135–144). Understanding that environment, including material culture, plays a parenting role in our psychological development is a prerequisite for establishing a solid foundation for any conceptualization and enactment of meaningful sustainability (Overmann, 2017).

Children are, of course, never *just* passive recipients of the actions of their parents or caregivers, but rather also actively mold the parent-child dyad (whether with a heart-melting smile or by inflicting prolonged sleep deprivation, for example). People(s) are similarly never simply and entirely passive recipients of the guidance and parenting provided by environments. If we have, as Crutzen and Stoermer (2000) have argued, entered the era of the Anthropocene, then Western civilization could certainly be considered as the planetary *enfant terrible* terrorizing their parents and interfering with their capacity to continue to provide good parenting. Environment, including material culture, has, in other words, been induced into a systemic practice of bad parenting and unsustainability. While this process is most pronounced in the current era of globalized neoliberalism, I argue its origins can be traced back to the agricultural revolution of the Neolithic.

Sustainability policy-development processes might also be understood in terms of this metaphor. Just as children victimized by especially poor parenting often struggle to recover from the psychological damage, at the scale of a society, those born into a world of unsustainability will necessarily struggle as well. Children who survive traumas, such as the Canadian residential school, Romanian orphanage, or the child soldier, slave, or child sex worker experiences, must deliberately seek out the traces of good-parenting skills and practices wherever and however they can be recovered. In extreme cases, such practices may need to be invented *ex nihilo*. This is not to say that environment necessarily needs some sort of (psycho-)therapy designed to make it a better parent, able to compensate for increasingly traumatized, stunted, and threadbare children. The nature of the child-parent relationship works against the efficacy of children curing parents. Successful environmental-policy development would, on the contrary, involve the establishment and institutionalization of processes that seek guidance from these residual traces rather than working to impose mastery upon the world. The idea of environmental "management," in other words, is a problematic image of the relationship if the goal is sustainability. We must not manage our parents, at least not before their condition truly requires palliative measures.

This shift in perspective, the framing of environment as acting like a parent, enables a new (or perhaps the rediscovery of a very old, nondualistic) understanding of our relationship with environment—a relationship that is dynamic and *action* oriented. Environment always guides us in a developmental process of co-becoming. The needed paradigm shift focuses on actions and processes that are always both material and mental, and must therefore first overcome the historical Cartesian cleaving of environment into either mind *or* matter. I am not alone in calling for this initial step of correcting "Descartes' error" (Damasio, 1994). Indeed, this book is intended to be indexical of, and to work to extend, a very broad and important paradigm shift that has been gathering momentum within the Western world for some decades now. That very recent shift can be recognized in the writings of many social and cultural theorists whose work, with a very few much earlier exceptions such as Giambattista Vico (1668–1744), dates mainly from the turn of the twentieth century. These theorists include Bateson (1972/2000), Boulding (1978), Burke (1984), Capra (1996), Damasio (1994), Fisher (2013), Fry (2012), Gell, (1998), Gosden, (1994), Haraway (2016), Heidegger (1962), Hoffmeyer (2009), Ingold (2013), Knappett (2005), Latour (2005), Macy and Brown (1998), Malafouris, (2013a), Mead (1934), Plumwood, (2002), Vygotsky (1978), Wertsch (1998), and countless other scholars whose work acts to break down and provide an escape from the dualistic assumption that mind and matter exist as two easily distinguishable and separate forms. This paradigm shift is observed, for example, by Stefanovic (2000), who states that

> it is no longer possible to conceive of human beings as merely individual subjects, primarily engaged in a Faustian manipulation of an infinite supply of environmental resources. Instead, we are compelled to come to terms with the fundamental, ontological truth of finitude, at the same time we acknowledge that our very being is defined by our immersion in-the-world and our concernful dealings with entities other than our own egos. (p. 143)

Every child needs to accept at some point that parents do have limits, and the role of the child is not to destroy the parent by being over-demanding. Smith (2011), also working with the conceptual tools of phenomenology, provides another example of the developing post-Cartesian perspective:

> The world is not a resource to be shared (divided out) among us; it is the appearance of the "between us" created through our interactions; our involvements; our resonances and dissonances; our collusions, conflicts, and concerns; our mutual exposure—our coappearance in/as the world. (p. 213)

The appearance of the "between us" is also the essence of the parent-child relationship, neither entirely whole without the other.

The break with the Cartesian paradigm almost certainly represents the most difficult and the most necessary challenge humanity has ever faced. The Western-Cartesian perspective currently forms the naturalized background and provides the inherited set of mediational means, guiding human action in our globalized world. My goal in what follows is to demonstrate that the challenge of sustainability essentially involves the transformation of our shared mediational means—that is, our cultural tools, those organized and organizing patterns, ways, and means, of getting on in the world. These tools mediate the interaction between person and world and thereby guide the co-development of both.

The types of tool kits I will be discussing are not limited to material artifacts, although these are certainly included and central to the argument to be unfolded. Indeed, meaningful engagement with stone tools may very well have provided the requisite evolutionary context for the emergence of the contemporary human being (Malafouris, 2013a, pp. 153–205; Renfrew, 2009, pp. 67–152; see also Chapter 2). The cognitive engagement with stone cutting tools in social groups provided, in other words, an evolutionary advantage to be able to participate in communities of practice and share a common set of mediational means. That said, conceptual and cognitive tools, such as languages, economic models, myths, metaphors, signs, icons, or even tacit (and subconscious) knowledges, are only a few more contemporary examples that still do not begin to exhaust the mediational means (or forms of parenting the world exercises on us) that, when put into practice, provide form, texture, speed, and direction to the processes of sociocultural as well as ecological development. That development also includes shared (culture-specific) aesthetic sensibilities and emotional practices and extends across the brain–body–tool set–environment spectrum. Our social and material environments teach and guide us to learn how to think and feel.

Mediational Means

Drawing upon earlier work by Vygotsky, Wertsch (1991, 1998) provides what is arguably the best framework for introducing the concept of mediational means, outlined first with an illustrative example, followed by a set of salient characteristics of mediated action. In *Voices of the Mind*, Wertsch states,

> To use a contemporary example, let us consider the case of an engineer who employs computer imaging to formulate options in the designing of the body of an automobile

> and to decide among them. Does it make any sense to isolate the mental action of the individual from the mechanism that mediates this action? Indeed, is it possible to say that such action is independent of the mediational means? (1991, p. 15)

Wertsch then elaborates in *Mind as Action*:

> In the view outlined here, almost all human action is mediated action. Given this, one would expect that it is very difficult to provide an exhaustive list of action forms and mediational means ....[However] I outline a set of basic claims that characterize mediated action and cultural tools ...(1) mediated action is characterized by an irreducible tension between agent and mediational means; (2) mediational means are material; (3) mediated action typically has multiple simultaneous goals; (4) mediated action is situated on one or more developmental paths; (5) mediational means constrain as well as enable action; (6) new mediational means transform mediated action; (7) the relationship of agents toward mediational means can be characterized in terms of mastery; (8) the relationship of agents toward mediational means can be characterized in terms of appropriation; (9) mediational means are often produced for reasons other than to facilitate mediated action; and (10) mediational means are associated with power and authority. (Wertsch, 1998, p. 25)

These mediational means, mediated actions, and cultural tools are an important locus of (un-) sustainability, although it is important to also recognize that (un-) sustainability spreads recursively and reflexively beyond mediational means and throughout the entire ecology of mind. The ecology of the mowed lawn (see Chapter 6) provides one good example of mediational means—the public display of mastery via mowing, watering, fertilizing, local bylaws, etc.—spilling over into processes of soil exhaustion to complicate, reinforce, and extend the initial compulsion. Such mediational means are especially important as a source and vehicle for the generation, transmission, distribution, and social enforcement of (un-) sustainability.

Mediational means are enlisted to guide the performance of almost any task including, in some circumstances, such seemingly automatic functions as being afraid (Davidson & Smith, 2003). The developmental process of absorbing and being absorbed into a set of culture-defined and culture-defining mediational means will also instill a set of values, for example, providing answers to questions: What is beautiful? or What is scary? At that point, those mediational means become the primary agent triggering reactions in the endocrine and cardiovascular systems when some culturally defined, beautiful, or fearful thing is suddenly encountered. The implication for sustainability is that all the routine behaviors that aggregated together result in unsustainability are all normalized through the mediational means we master and that, in turn, master us.

The Western-Cartesian model of subjectivity, a subjectivity derived from a very specific tool set of inherited mediational means, is transmitted from generation to generation primarily within the vehicle of those same mediational means as an integral component of a complex entirety of historical, sociocultural-ecological inheritance—the already-lived-in world, into which we are thrust from the moment of conception. The fetus's perceptions of the world, whether transmitted chemically or hormonally from the mother or in the form of in utero soundscape, begins from the start to be etched into the fabric of the person–cultural tool–environment triad. The Western-Cartesian already-lived-in world has been filled to overflowing with traces of unsustainability that get smuggled into us through the mediational means we inherit and inhabit whether we think it does or not. Often these influences are characterized as ideas and narratives. Plumwood (2002), for example, refers to the

> "historical rationalist imaginary" figured through Cartesian dualistic assumptions that determines that women and other "lesser beings" are the Others of reason, which is treated as the province of elite men who are above the base material sphere of daily life and are entitled to transcend it because of their greater share in Reason. It is not only women that have been constructed as oppositional to western rationality, culture and philosophy, but also the slave, the animal, and the barbarian, all associated with the body and the whole contrasted sphere of physicality and materiality. (p. 19)

In Wertsch's (1998) terms, we can see in Plumwood's "historical rationalist imaginary" an "irreducible tension between agent and mediational means" (p. 25). People are mastered by their tools as much as they implicitly master their roles and identities through the "situated learning" that arises from putting such tools into "communities of practice" (Lave & Wenger, 1991, 98–100). However, these influences are not limited to, or simply caused by, a historical imaginary but are also and always mediated through material artifacts, landscape configurations, and other mediational means. Women and other beings are constituted as lesser within a *social mind* that is always a group of individual minds coordinated with a set of shared mediational means including assemblages formed as modes of dress, etiquettes, accepted practices, social norms, economic institutions, architectures, designs, and so on. There is always a material dimension, whether involving material artifacts or not. Even the choice of words spoken or the body language these tools call forth are also materializations, even if they were only to manifest in the form of the sound waves of speech or the synaptic firing through the human nervous system. Mediational means always involve consequent and selected forms of synaptic firing in brains and consequent and selected material actions in the world, crossing the illusory Cartesian mind-matter divide.

Plumwood's (2002) "historical rationalist imaginary" becomes evident more as a world-forming pattern than it does in individual instances of denigrating "lesser beings" (p. 19) or prioritizing the mental over the material. Clearly, meaning emerges from the practice of treating "women and other lesser beings" as such and not through the simple application of thought or reason to a separate materiality. But even this understanding of practice frames it as something applied to the world and suggests that all that is required to change this world is within the purview of our conscious choices to act in certain ways. But things are more complicated. For one thing, putting any type of cultural tool into practice is seldom a straightforward or obvious endeavor and we often cannot even say why we do the things we do. Polanyi (1966, p. 1) referred to this phenomenon as "tacit knowledge." What is more, each time a specific cultural tool is put into practice, a particular way of becoming in the world becomes more concrete and other potentialities simultaneously fade into abstraction amid the traces of lost opportunities. According to Malafouris (2013a),

> Cultural knowledge and innovation are not intracranial processes; they are, rather, infused and diffused into settings of practical activity, and thus they are constituted by experience within these settings through the development of specific sensibilities and dispositions, leading people to orient and think about themselves and their environment in specific and often unexpected ways. (p. 116)

It might seem strange to describe the institutionalization of women, animals, and barbarians as lesser beings as "cultural knowledge" or "innovation," but all social change occurs through an innovation of mediational means achieved through practice. The perspective of mediated action, freed from the illusions of positivism and "calculative thought" (Stefanovic, 2000, p. 28), provides a more realistic analysis that leans toward a constructivist or phenomenologist perspective without absolutes, only the relational metrics of better and worse. In all of this, we need to recall that in trying to elucidate these mediational means, there is nowhere to stand outside of these mediational means, no view from nowhere (Beatley, 2005; Cosgrove, 2001) but, rather, only a partial view, a partial understanding, and a mediated experience, always situated within and through a culture-specific historical inheritance applied through the enactment and continuation of an already-lived-in world.

In what follows I will generalize by referring to the entirety of contemporary unsustainability mediational means, including that identified by Plumwood (2002) as the "historical rationalist imaginary," (p. 19) as the "Western-Cartesian paradigm/worldview" or "contemporary conventional wisdom." The Western-Cartesian paradigm is a rather slippery concept, but for the purposes of this book, it includes both learned and implicit conceptualizations of the way the world works. It could be thought of as one specific set of mediational means, a collection of tools used to form patterns of perception, thought, and emotional practices that are far from being simply a result of Descartes's philosophy. Indeed, I argue that they originate as far back as the Neolithic origins of Western societies in the fertile crescent when any given space began to be categorized into domains of either the home or the not-home, the safe and controlled, or the unfamiliar and dangerous. This expanded gradually into a dualistic system of sacred/profane, safe/unsafe, cultivated/wild, male/female, mind/matter, and so on (see Enz, 1976). Today, these mediational means have become as ubiquitously inevitable as anthropogenic toxic-chemical compounds in our air, water, and soils and, I will argue, this Cartesian ontology plays the key role in the propagation and ecological-sociocultural maintenance of contemporary unsustainability (see Gifford & Gifford, 2016; Varey, 2011).

Interest in shifting our paradigm away from the Cartesian model to provide an alternative way of becoming in the world is evidenced by, for example, newfound interest in and respect for Indigenous knowledges, sustainable design theory, Transition Towns, land trusts, community gardens, guerilla gardening, dumpster diving, and many other manifestations of people cultivating new place-based identities more consistent with an environment-as-parent perspective. This shift can no longer be entirely ignored, and indeed *should* no longer be resisted, given the current ecological crisis.

What these post-Cartesian scholars and practices all share is not primarily a focus and concern with the establishment of a re-engineered ethics, but an implicit turning away from an obviously flawed (I would even say delusional) model of the way the world works and toward a more relational epistemology and ontology, including a focus on materialized knowledges. In some ways, this movement, as expressed within the social sciences and humanities, extends what has already taken place in physics, where Einsteinian (or quantum-physics) perspectives supersede and replace the increasingly inadequate Newtonian understandings and frameworks (Capra, 1975/2010).

This book aims to contribute to this paradigm shift—that is, to recognize that the current unsustainability is joined at the hip with a specific way of becoming in the world that is, on the one hand, historically anomalous, while, on the other hand, still globally predominant. It critiques various examples of Cartesian ways of understanding, feeling, embodying, enacting, and materializing the old paradigm, including, for example, mainstream economics, mowed lawns, and Diamond's (1999) answer to "Yali's question" in *Guns, Germs, and Steel: The Fate of Human Societies*. It also seeks to provide a theoretical framework that can offer important new insights into humanity's greatest contemporary challenge, that is, (un-) sustainability (Barry, 2012).

Yali's Question

Yali's question, from the opening pages of Diamond's (1999) *Guns, Germs, and Steel* provides a convenient and insightful example with which to begin. Yali asked Diamond: "Why is it that you white people developed so much cargo and brought it to New Guinea, but we black people had little cargo of our own?" (Diamond, 1999, p. 4). The question is in many ways the same as asking: Why are some societies sustainable and others not? Diamond's reply focuses, rightly, on what constitutes the difference between two very different cultures with two very distinct person–mediational means–world assemblages. Unfortunately, Diamond's answer entirely misses the part of the story most pertinent to contemporary collective efforts to achieve sustainability. The four-part answer put forward by Diamond (1999) was that there were differences in Western and non-Western environments:

> differences in the wild plant and animal species available as starting material for domestication ...[differences in] rates of diffusion and migration .... They were most rapid in Eurasia, because of its east-west major axis and its relatively modest ecological and geographical barriers. Diffusion was slower in Africa and especially in the Americas, because of these continents' north-south major axis and geographical and ecological barriers .... Related to these factors affecting diffusion *within* continents is a third set of factors influencing diffusion *between* continents, which may also help build up a local pool of domesticates and technology .... The fourth and last set of factors consists of continental differences in area or total population size. A larger area or population means more potential inventors, more competing societies, more innovations available to adopt—and more pressure to adopt and retain innovations,

> because societies failing to do so will tend to be eliminated by competing societies. (pp. 405–408)

Diamond's answer would have to be rather different if it considered the bodies of literature, exemplified by Wertsch, Bateson, and Malafouris's Material Engagement Theory (covered in the next chapter) that address, problematize, and clarify the ways matter and mind together form a unified cognitive system. Though Diamond looks to environment as a key explanatory term, he does so in a physically determinist manner, while also ignoring the mediated co-constituting agency of environment for psychological development.

By misunderstanding the way "matter matters" and the way "stories make worlds and worlds make stories" (Haraway, 2016, p. 12), Diamond unwittingly advances a Cartesian account of the way the world works. The structures and processes described in *Guns, Germs, and Steel* and the way they emerge and spread as cultural practices is itself iconic of one specific and historically anomalous mediating sociocultural/cognitive tool set. Notice the implicit assumption that we are in an evolutionary race with our neighbors to achieve "progress" as quickly as possible to avoid being eliminated. Notice the positivist assumption that scientific knowledge and technological innovations are cumulative in nature and always represent a progressive linear, non-reversible process. Diamond presents a recursive history where there was only ever one environmentally determined answer as to who would win the race.

Although Diamond's thesis has been critiqued from many directions, and not least by focusing on the many exceptions to the rules that Diamond attempts to establish, the broad strokes of his narrative continue to resonate with a society where there is widespread acceptance of there being "one right way" (Séville, 2017, p. 449), only one possible world, (or, as articulated by former British Prime Minister Thatcher, "there is no alternative"—TINA). Neoliberalism and scientific positivism both assume a similar model of increasing economic and epistemic mastery through increasingly sophisticated manipulation of the world. In this sense, at least, the Diamond thesis operates within an ecology of mind that is representative of a much broader unsustainability.

Diamond's perspective overlooks how, for example, the invasion of Turtle Island by the Europeans primarily concerned the confrontation of two very different social minds—two different cultural tools sets resulting in two different ways of perceiving, cognizing, feeling, knowing, making, and interacting with the world; two very different material cultures; two different ontological realities; and two very different mediated ways of becoming in the world that meant two very different consequent human worlds and two very different world trajectories (Richter, 2001). The difference, though, is not limited to ideas but rather extends into such material structures as "the opaque workings of finance capital; the harsh machinery of social security, juridical and penal systems; and the ideological pressures of educational institutions, the media and the internet," (Mishra, 2017, p. 111) brought about by one-right-way approaches.

The Indigenous population and the European settlers represented two distinct categories of ways of becoming in the world, including different ways of relating to the world—readily distinguishable and identifiable, at least in part, by the relative rigidity/flexibility of the respective person-tool-environment triad constituting cultural tool sets. If, as Mishra argues, one-right-way neoliberalism has set us on a road that refuses to recognize how the wheels are falling off the bus or take anything resembling an exit ramp, many other knowledge systems have been able to accommodate radical change (Lear, 2006) while maintaining a familiar cultural identity. We could think of a theoretical continuum with societies most able to accommodate changes to their way of life at one end and societies with built-in mechanisms of resistance to any fundamental change on the other. The relative positioning on a theoretical rigidity/flexibility continuum is largely determined by those societies' mediational means.

The mediational means of many Indigenous knowledge perspectives are context specific. Rather than tending to frame any knowledge or tool as being the universal "one right way," Indigenous knowledges, like Indigenous parenting practices, tend to be somewhat flexible, malleable, with lots of alternatives, whereas Western knowledge aspires to solidity, finality, objectivity, and the elimination of alternatives. Newhouse (2004), for example, states,

> Indigenous scholarship also brings to the table the notion of "sense- making," the collective coming to understand something. Sensemaking is rooted in identity. Making sense becomes a collective process. Let us put our minds together, let us come to one mind, let us find consensus—these are examples of the discourse that we see around us. Indigenous scholarship offers the promise of collective knowing, rather than expert knowing. (p. 144)

The process of sensemaking and collective knowing is quite different from the process prescribed by the so-called scientific method, but even more striking are the differences in the products of those two types of processes.

Diamond's thesis, for example, includes the explanation of European superiority in food production vis-à-vis the Americas, based partly on a difference in the number of large mammal species amenable to domestication. Diamond's focus on domesticated animals is characteristic of a way of seeing, feeling, knowing, and becoming that is more notable for what it forecloses than for what it enables. Notice how well Diamond demonstrates mastery of Plumwood's (2002, p. 19) "modern rationalist narrative of domination of the Others." According to Diamond (1999),

> Eurasia's 13 species ... became its chief source of animal protein (meat and milk), wool and hides, its main mode of land transport of people and goods, its indispensable vehicles of warfare, and (by drawing plows and providing manure) a big enhancer of crop production. Until waterwheels and windmills began to replace Eurasia's mammals in medieval times, they were also the major source of its "industrial" power beyond human muscle power—for example, for turning grindstones and operating water lifts. In contrast, the Americas had only one species of big domestic animal, the llama/alpaca, confined to a small area of the Andes. While it was used for meat, wool, hides, and goods transport, it never yielded milk for human consumption, never bore a rider, never pulled a cart or plow, and never served as a power source or vehicle for warfare. (p. 355)

Notice especially Diamond's implicit definition and understanding of the concept of food-production success. Control, or domestication, figures large in this imaginary. Domestication equates to security of accessibility and speaks to an underlying assumption of scarcity, putting the background tone on a setting of *anxiety* that colors every aspect of Western existence. The type of relationship one can establish and maintain with the world is very different if that world is assumed to be miserly than if that world is assumed to be one of generosity and plenitude. These different assumptions form a key aspect of the defining mediational means that distinguish one culture from another and where those worlds are situated with regards to their comparative rigidity/flexibility and (un) sustainability.

Richter (2001) puts forward a template very different from the historical rationalist imaginary and enables a different understanding, assessment, and accounting of Indigenous food production:

> The contrasting modes of interaction between humans and their environment are illustrated by the ways in which Native and European agriculturalists used the land. Perfected relatively recently during the Medieval Optimum and honed in the harsher conditions of the Little Ice Age, Indian agriculture was supremely adapted to the

> eastern North American environment. *Its productivity put its European counterpart to shame*. (p. 55; emphasis added)

The Indigenous mind encountered by the European invaders was not simply more agriculturally productive; it preserved and enhanced food production *capacity*—the stock of healthy soil and biogenetic diversity that, in turn, determined the maximum potential net primary production (NPP) or food production for a given place over the long term.

This ecologically sustainable culture-environment balance may not have always been the case. Like Australia, the Americas experienced the extinction of most of its megafauna following the arrival of humans. We cannot be sure how many thousands of years it may have taken (a luxury no longer available), but in both the Australian and American historical experiences, the Indigenous cultures did eventually find ways to ultimately achieve a vastly superior version of sustainability than their Western-Cartesian counterparts. Western "civilization" had already greatly degraded the ecological capacity of the Fertile Crescent, the entire watershed of the Mediterranean Sea, and much of Europe by the time of the Turtle Island invasion. This was accomplished without any apparent ecological learning whatsoever. The systematic deforestation and ecological degradation experienced in the originating countries of the Western-Cartesian paradigm usually failed to make any compensating adjustment to that persons–mediational means–world system. These same mediational means continued this pattern of deforestation, soil erosion, and collapse of ecological productivity (Cronon, 1983) when introduced to Turtle Island, with rapid and irreversible effects. As soon as materiality could be configured in toxic ways, it was, for example, in the forms of pesticides, plastics, concretes, asphalts, and psychologically damaging architectures.

Why have the ecological consequences not resulted in ecological learning outside of a few notable exceptions such as disciplines like historical ecology and sustainable design, or environmentally minded groups and individuals? This systemic oversight might partly be understood as the result of an ontological blind spot created by the delusional tools of the European social mind. Among those very few early invaders who did notice the apparent inferiority of the European way was Marc Lescarbot, a passenger on the *Jonas*. "Captained by Samuel de Champlain, the *Jonas* was on a mission to resupply New France, presently known as 'Quebec,' in 1604." (Todd, 2001 p. 12) Todd quotes Lescarbot's diary entry: "We had no other domestic animals save hens and pigeons." (Ibid.) By contrast, Lescarbot

> chronicled a New France bursting with life overlooked by his gold and fur-seeking peers: lobsters, mussels, cod; thick forests and raspberries; eagles so plentiful as to be a nuisance … feasting on the settlers plump birds as an imported delicacy. (Todd, 2001, p. 12)

Where I currently reside, on the Island of Montreal in southern Quebec, there would have been an almost unimaginable quantity of fish and game available at the time of Lescarbot's voyage on the *Jonas*, without any need of domestication. Gone now are the vast communities of salmon, sturgeon, trout, turkeys, grouse, deer, rabbits, and other fish and game that kept Indigenous peoples healthy in body and mind, as triadic person–mediational means–environment assemblages. Western agriculture failed to compensate for the increasing absence of fish and game:

> Annual eastern Indian corn yields may have been nearly nineteen bushels per acre. Although that return was roughly equal to what European cereals provided under similar environmental conditions in the same period, it was achieved with far fewer plants: one kernel of maize might return as many as two hundred kernels at harvest, yet contemporary wheat yielded only about fifteen grains for each seed. A Western European acre thus had to be densely packed with a single crop, but a North American acre had plenty of room left for beans and squash as well as corn. (Richter, 2001, p. 55)

These plants and animals nourished and kept healthy the Indigenous body, brain, culture, and environment. Their loss cannot be adequately measured using simple quantitative caloric, nutritional, or monetary metrics.

This is not to suggest all the Indigenous inhabitants of Turtle Island always experienced a continual surplus of food and resources and a material abundance vastly superior to that offered by Europe. There may also have been examples amongst Indigenous peoples of the Americas, possibly rare and isolated, of ritualistic practices of ecological and psychological destruction often associated with unsustainable development trajectories (such as those noted by Diamond regarding the Maya Empire). But a comparison between North American cultures and their mediational means, of "actual minds and possible worlds" (Bruner, 1986) is still immensely valuable. Benedict (1934) offered these words as the growing threat of World War II gathered on the horizon, a threat not entirely unlike that of the ecological catastrophe facing us today:

> Social thinking at the present time has no more important task before it than that of taking adequate account of cultural relativity …. The sophisticated modern temper has made of social relativity, even in the small area which it has recognized, a doctrine

> of despair. It has pointed out its incongruity with the orthodox dreams of permanence and ideality and with the individual's illusions of autonomy. It has argued that if human experience must give up these, the nutshell of existence is empty. But the recognition of cultural relativity carries with it its own values, which need not be those of absolutist philosophies. It challenges customary opinions and causes those who have been bred to them acute discomfort. It rouses pessimism because it throws old formulas into confusion, not because it contains anything intrinsically difficult. As soon as the new opinion is embraced as customary belief, it will be another trusted bulwark of the good life. We shall arrive then at a more realistic social faith, accepting as grounds of hope and as new bases for tolerance the coexisting and equally valid patterns of life which mankind has created for itself from the raw materials of existence. (p. 287)

Perhaps any historical cultural inheritance will come to be taken as the best way to be in the world if it gets etched into the fiber of one's being, but there are always possible alternatives. Three decades after Benedict, Sahlins (1966) described some of the last remaining hunting and gathering peoples on earth as "affluent" and offered much the same advice as Benedict, but so far, to no avail. I will be arguing, over the following pages, that by examining different food-provisioning regimes, different aesthetic sensibilities, and different embodiments and materializations of different ways of relating to and becoming in the world, we can gain important insights into the underlying suite of mediational means that ultimately holds (un-) sustainability in place.

The sets of mediational means co-constituting persons into one cultural category or another do not just show up in ways of thinking, though. Sustainability is not strictly a knowledge-based problem but a problem of mental-material systems (ecologies of mind) creating landscapes of ecological consequence. The problem lies in the materialization of mediational means, which is why, in Haraway's (2016) words,

> it matters what matter we use to think other matters with; it matters what stories we tell to tell other stories with; it matters what knots knot knots; what thoughts think thoughts; what descriptions describe descriptions; what ties tie ties. It matters what stories make worlds, what worlds make stories. (p. 12)

The story told here is that of (un-) sustainability in relation to matter mattering—stories told to tell other stories, stories making material worlds, and worlds making stories. It focuses on the dominant, the obscure, and the potential sociocultural models of subjectivity, shared and collectively internalized from places into bodies and brains and back to places in unbroken circuits. It focuses on how the inherited and internalized models of subjectivity, the readily available sets of mediators of sociocultural action (Wertsch, 1998), derived from collectively routinized practice in the already-lived-in world, largely constitute individual minds in similar ways through the forms and frequencies of those material-engagement processes (Malafouris, 2013). It especially focuses on how these readily available tool sets are shared by groups of individual minds to form an entire ecology of regularized thought and feeling to constitute a *social mind* (Clark, 2002; Forgas, Williams, & Wheeler, 2001; Valsiner & van der Veer, 2000; Zerubavel, 1997, 2015).

The inherited and self-mastering Western-Cartesian perspective deceives us into thinking we are someone we are not. The person and people we are, are in fact inseparable from our cultural and material developmental context. We are inseparable from our mediational means and from the environments that parent us. This story, I suggest, will provide a better answer to Yali's question.

Mediated (Un-) Sustainability

Sustainability is often, correctly, framed as an issue of ethics. However, a problem arises if the mediational means, the cultural tool set available for Western peoples to construct an ethics of sustainability, is simply not up to the job. The dominant tendency might be to try to outline a specific ethics for sustainability to be applied universally and "formulaically to ecological concerns" (Smith, 2011, p. 48). But this seems to assume that such an ethic could be formulated in accordance to the way the world "really" is rather than recognizing that we experience and understand the world through an inherited, and somewhat historically arbitrary, set of mediational means, a tool set that also includes those same philosophical presuppositions and approaches. Embedding ethics in abstract principles and calculations is a problem that certainly pre-dates Descartes (see Lopes, 2017, p. 179; Smith, 2011, pp. xi–25), but the point here is that the contemporary tools we have for thinking about ethics cannot get us to sustainability any more than quicksand can provide a good foundation for a house. Perhaps the best we can do is "rough out" values that will often escape moral formulations that are, in that sense, pre-ethical.

Here parenting might again provide an apt metaphor. Parenting is replete with ethical implications, but the child-parent dyad does not easily fit into Western conceptualizations and theoretical frameworks of rational and/or sovereign actors. Here, too, any attempt to provide abstract principles only gains social currency insofar as those principles resonate with the pre-existing cultural context, a context that includes an economic system organized around the principle of self-centered individualism. This would not bring us very far in the direction of sustainability, to say nothing of good parenting for any human child.

Bateson (1972/2000; see also Bateson and Bateson, 1987) recognized that for sustainability to be successful, it would have to be "sunk" to the level of the subconscious, a place Bateson referred to as "the sacred" (1987, pp. 80–81). This is also echoed to some extent by Stefanovic (2000), who states,

> There is a fundamental moment of discernment in the moral decision making process that distinguishes it from the idealized version of technical knowledge that we possess in a modern, scientific, technological era. Moral knowing is a special kind of knowing, to be distinguished from technical knowledge to the extent that the former involves not only a manipulation of explicit facts, but a unique seeing and attunement to the moral imperative. In Gendlin's words, ethics must, consequently, involve both inclination as well as concepts. (p. 126)

Here too, though, we need to recognize that the sacred and our inclinations, as well as our concepts, reside in an always mediated, but also situated, extended, and distributed social mind. "Moral decision making," in other words, never occurs exclusively inside a computer-like brain but rather takes place collaboratively between a child-like agent and a parent-like world, as Malafouris (2015) explains:

> The relation between brains, bodies and things (in the broad sense of the relevant material environment, natural or artificial) is not one of representation, not even one of interaction. Instead it is a transactional process of mutual constitution. It is only by understanding the different forms and properties of this transactional co-constitution that we will ever be able to understand the remarkable plasticity of the human mind. (p. 354)

The development of any ethical inclinations and/or concepts needs, in other words, to involve the extended mind, including brain, body, cultural tool sets (both mental and material), and environment. It needs to be established across all these domains as a ground upon which sustainability as a moral practice *could* be established.

I want to focus, then, on the requisite platform upon which any hypothetical ethical system could even be contemplated or tentatively constructed. Initially from birth, those foundational patterns of sociocultural historical inheritances provide us with a platform upon which to establish all our cultural practices and then mediate any ongoing processes of enactment, embodiment, and materialization of (un-) sustainability. This is not an inheritance in the usual way we think of it, whether material, narrative, or biogenetic. Rather this inheritance involves an entire world: the cultural landscapes we inhabit including the entirety of material artifacts, and the cultural and cognitive tools we use to structure our practices of perception, emotion, and cognition—both conscious and subconscious, as well as the entire historical unfolding of life in all its vast diversity. In short, this inheritance comprises the entirety of the circuits of thinking, feeling, making, and becoming running between us and any given aspect of the already-lived-in world.

For the human mind to operate as the tool-making and tool-constituted species, that entire world must be brought into a process of collaboration or partnership so that matter offers itself up in the form of the cultural-cognitive mediator actively participating and making contributions to an implicit, mostly subconscious project with specific (un-) sustainability characteristics. The point is not simply that humans *require* tools, but rather that the tools humans require, and the tools that they are then necessarily mastered by, transform the human experience of being in the world into a specific situated and mediated way of being. We are, as Clark (2002) rightly claims, "*natural-born* cyborgs" (p. 3) and parent-like environments have constituted us as such.

This links us back to Wertsch's (1998) development of Vygotsky's (1978) work (see above). As Cole and Scribner (1978) explain,

> Vygotsky brilliantly extended this concept of mediation in human-environment interaction to the use of signs as well as tools. Like tool systems, sign systems (language, writing, number systems) are created by societies over the course of human history and change with the form of society and level of its cultural development. Vygotsky believed that the internalization of culturally produced sign systems brings about behavioral transformations and forms the bridge between early and later forms of individual development. Thus for Vygotsky, in the tradition of Marx and Engels, the mechanism of individual developmental change is rooted in society and culture. (p. 7)

It should be noted that there are normal stages of human development that do not necessarily occur in a linear fashion. Cultures likewise do not necessarily develop in any specific direction, although the change process itself can recursively appear to occur in a step-like fashion, similar in some respects to stages of ecological succession or the stages of psychological development in humans.

Wertsch (1998), too, adds an important caveat to Vygotsky's insight on mediated action arguing that his preferred terms of

> "mastery" and "knowing how" have some advantages over the more general notion of internalization in several respects. The first of these is that it makes it possible to avoid some unneeded conceptual baggage that comes with the term "internalization." Many forms of mediated action are, and indeed must be, carried out externally. It may not be a necessary entailment of the term, but "internalization" suggests an image in which processes that were once carried out on an external plane come to be executed out of sight on some kind of internal plane. This is the kind of image encouraged by discussions such as those by Vygotsky (1978) about how counting originally occurs on an external plane with the help of material cultural tools such as sticks or one's fingers and then disappears as it is internalized. (p. 50)

Mediational means, as noted above, are not always located "inside" the person; even when they are, they tend to leak, so the term *internalization*, although useful for Vygotsky, is best replaced with terms denoting the bringing of mediational means into a community of practice. Wertsch's own favored term *mastery* is also somewhat dubious since mastery is also exercised by the tools over their users, as Franklin (1991), for example, makes clear in her discussion of "prescriptive technologies." In terms of formulating any ethics, this further complicates matters. Must we then consider external material cultural tools such as sticks or cars or video lottery terminals as legitimate subjects for ethical theorizing? (See Carter & Palermos, 2016.) The key point here, though, is that, as Wertsch (1998) advises us,

> the essence of examining agent and cultural tools in mediated action is to examine them *as they interact*. Any attempt to reduce the account to one or the other of these elements runs the risk of destroying the phenomenon under observation. (p. 25; emphasis added)

The ethical "subject" then would be, for example, a person-and-stick or person-and-number system and the problem of ethics would need to be extended accordingly, especially since the "person" does not always choose to become "person-and-stick." We do not generally choose the language we learn from birth—the knowledges, conventional wisdoms, paradigms, myths, and metaphors—nor do we invent the material artifacts or cultural landscapes we find available to give organizing structure to our perceptions, thoughts, and feelings. Inclinations, as much as concepts, are mastered and mastering.

Once again, we often see analyses that recognize either that material environments matter *or* that ideas matter—that, for example, engage in the conceptual separation and prioritization of cognitive action (e.g., "values and ideologies" Dale, 2001, p. 100) over conceptualized as inert matter. According to Dale,

> The sophistication of modern society and its organizational structures are supported by inherent and interlocking values and ideologies. These, in turn, are tightly coupled with structural barriers that systematically reduce the ability and, indeed, the capacity of new concepts and alternative models to challenge dominant paradigms. Often these interlocking values and ideologies are shared across institutions and between sectors. Over time, this gridlock produces an overwhelming inability to respond, even in the face of new information and facts that illustrate the importance of immediate action. (p. 100)

This dominant view of "interlocking values and ideologies" in some sense causing "an overwhelming inability to respond" is certainly plausible. Certainly attitudes, values, ideologies, narratives, and other mediating tools, including the structure of language (see Bateson in Reiber, 1989), are used to provide organizing structure to our thoughts and emotions and do work to anchor regimes of (un-) sustainability. But if we fail to recognize the role materiality plays in social and individual cognitive processes, the challenge of (un-) sustainability simply remains incomprehensible, locked inside an impenetrable black box. Values and ideologies are always, in other words, interlocked with and through material cultural artifacts and physical-landscape features. These form the stable materializations of those values and ideologies of the already-lived-in world. In short, matter matters, too.

The inheritance of cultural tools is transmitted through normal human socio-psychological-development processes and that transmission and ecological developmental process continues throughout the entire lifespan (and even beyond, through the remnants of invented tools and changed constellations, forms, and patterns of the many traces that we leave behind even after death). Bateson (1972/2000) uses the language of systems theory to make a similar point "In no system which shows mental characteristics can any part have unilateral control over the whole. In other words, *the mental characteristics of the system are immanent, not in some part, but in the system as a whole.*" (p. 316; italics in original). This point has also been taken to heart within Material Engagement Theory, building on the earlier concept developed in the context of philosophies of (materially) extended mind put forward by Clark and Chalmers (1998) known as the "parity principle" (p. 8):

> If, as we confront some task, a part of the world functions as a process which, were it to go on in the head, we would have no hesitation in accepting as part of the cognitive process, then that part of the world is (for that time) part of the cognitive process. (p. 8)

A Batesonian ecology of mind that includes the parity principle will be very different from a system constructed upon Cartesian dualistic assumptions. Ideas and conceptual frameworks form an integral part of the sociocultural-ecological system, but only a part. Similarly, many material artifacts might find a role within situated, extended, and distributed minds, being put into practices of sustainability or unsustainability. Once those practices are established with those various constituents of brain-body-mediational means-environment assemblages, it can become extremely difficult to break and re-form those chains.

Some of these chains may extend far back in time. For example, "In early literary history" according to Harrison (1992), "forests begin to appear early on the scene for what later comes to be known as the 'unconscious'" (p. 87). This early use of the forest as a metaphor for the unconscious can tell us a lot about the organizing narratives people used to inform both their actions and the resultant material and conceptual artifacts—the various traces that formed the already-lived-in world for each succeeding generation, a succession we might refer to as a loop of anxiety and fear of the Other.

A learned, reality-based, fear of the unknown or uncontrolled has been a dominant theme throughout the entire history of Western civilization as Winston (1997) explains:

> Beginning around 10,000 years ago, with the advent of agriculture and denser human settlements, our relationship with pests began to change. Increasing urban populations, compact crop plantings in fields and orchards, herds of domesticated animals, and stores of grain, vegetables, cloth fibers, furs, and dried meats all provided concentrated food sources for organisms that previously foraged widely for sparse food sources. In addition, we began to disrupt entire ecosystems and their inhabitants, transforming diverse natural habitats into cultivated, single-cropped fields and dense, sprawling cities. These changes induced the populations of a small number of species to explode into pest status. Trade added to this potent brew by transporting plants and animals far out of their natural ranges, to new habitats with exciting food sources and few predators, parasites, or diseases to keep them in check. (pp. 3–4)

The enactment and materialization of the conceptual separation between cultivated and uncultivated, or sacred and profane (Shiner, 1972), the safety of the home juxtaposed against the insecurity of the wild confronted ecological systems to form a sociocultural-ecological positive-feedback loop. In this loop, ritualistic segregation and compartmentalization led to catastrophe in the form of induced pestilence and disease, possibly followed by increasingly technologized violence as a solution to the consequent food insecurity and, eventually, increasingly dysfunctional political or religious institutions, from "big man" (Sahlins, 1963) to fascist dictator. In short, the type of insanity originating in a specific set of unsustainability mediational means began to be materialized into environment, making themselves available as an increasingly dysfunctional tool set for further mastery extending in both directions from person-tool to environment and environment to person-tool. Catastrophe, in turn, was experienced as a socially shared psychological trauma that resulted in increasingly dualistic processes, structures, and cultural-cognitive tools. The tools emerging from the experience of trauma, in other words, guided behaviors that would set up the next round of trauma.

The loop of fear and anxiety that characterizes contemporary mediational means continues to this day and is evidenced in many forms: symbolic enactments (treatments) in response to a variety of medicalized conditions including cancer, attention-deficit disorder, post-traumatic stress disorder (PTSD), sore throats, or ear infections; modern approaches to education conceptualized as the eradication of ignorance; counterproductive over-reliance on pesticides, prison systems, manicured suburban lawns, anti-immigration policies, and pre-emptive assassinations; and continuous and compulsive military adventures, as well as countless other routine daily examples. The current covid-19 crisis, although anthropogenic, tends to be framed in militaristic terms as an evil enemy fitting into this same ongoing pattern. That subconscious *pattern* guiding and normalizing the social-psychological neurosis of eradication of contagion—whether perceived as forest, disease, pest, ignorance, bad genes, radicalized enemy combatant, refugee, germ, or weed—serves as a convenient entry point for understanding the parameters and processes of the dysfunctional inheritance of unsustainability. These subconscious patterns act as epistemological and ontological default settings. They guide our psychological development from birth toward a specific way of becoming in the world that forecloses other potential settings; ultimately, they paradoxically "recreate reality" into a waking hellish nightmare of biodiversity collapse, plasticization of the planet, global climate change, food insecurity, global pandemic, and, in a word, "futurelessness." This cycle of configuring fear of contagion into the physical environment through ecological incompetence (say, mowed lawns or overprescribed antibiotics) as well as into ideological, cultural, and cognitive tools (the governmentality that naturalizes lawn-care and over-prescription), which then provide increasing evidence of and justification for the need to be afraid (as lawns and antibiotics fall prey to weeds and antibiotic resistance), forms the foundations of the contemporary ongoing practice and materialization of unsustainability.

The situation, then, might seem hopeless. But laying out this process and its relation to (un-) sustainability might take up the baton offered from Bruner (1986), who recognized early on that the pathway out of the contemporary "deep malaise, a malaise of futurelessness" was in our greater understanding, both conscious and subconscious, and skillful practice "to recreate reality, to reinvent culture" (p. 149). This reinvention of culture and reality involves the transformation of minds, co-constituted from brains and bodies, but equally importantly from tools and places.

Positivism

Although the term *positivism* has been employed in many ways, it might be used for current purposes to denote a set of cultural tools that suggest the objectivity of scientific practices and the epistemologically progressive abilities of this approach to accessing and "knowing" reality. In this sense, it might serve to illustrate the operation of aspects of heritable models of subjectivity that result in unsustainability. I will argue that contrary to positivism's tenets, the evolving inheritance of mediational means as an ongoing process never guarantees, of itself, any real improvement.

Although positivism forms the background context for almost all Western scientific-knowledge-production activities, it simultaneously and systematically takes that same manufactured knowledge out of context. The elements on the periodic table, the Law of Gravity, penicillin, and atomic bombs all tend to frame people as outsiders looking in. In other words, the reflexive nature of knowledge production as well as the ignorance that is always co-produced is generally ignored. Once water is known as a molecule composed of hydrogen and oxygen atoms, then some other ways of knowing water become more difficult. The mastery of one set of mediational means forecloses the mastery of another.

Wisdom or ethical action is therefore not the same thing as knowledge. Knowledge is something we make with one set of mediational means, but wisdom and ethical action is generally embedded within and grows with a set of mediational means. Knowledge can evolve rapidly, but wisdom, on the other hand, involves the slow evolution of a form of subjectivity. Positivism obscures our awareness and understanding of the speed at which, and the direction in which, we are traveling. Positivism removes the characteristics of our more-or-less predictable destination from legitimate debate. Worse, such an unstable foundation makes discussions of ethics more abstract and unrealistic. Today, for example, we inhabit a place where anything that *can* be invented is conceptualized as something that *should* be produced and marketed.

To the extent that positivism becomes materialized as a defining characteristic of the already-lived-in world, the normal human developmental process becomes one of being socialized into these Western-Cartesian mediational means. Unfortunately, an increasing correspondence and synchronicity (Knappett, 2005, p. 61, uses the more elegantly precise terms *isomorphy*, *tuning*, and *resonance*) between meaning system and material artifacts, whether at individual or societal scales, may in fact bring about increasingly negative outcomes if the material culture and physical landscape already contain psychotic levels of unsustainability. Mind holds no natural dominion over matter. Robbins (2007) provides an excellent example with a discussion of the mowed lawn that we will return to in detail later (see Chapter 6):

> That the needs of the lawn are interpreted through human cognition, culture, and meaning-making does little to change the urgency of the lawn's impressions on the people who do lawn work. Indeed, the taken-for-granted obviousness of the lawn's needs is one of the things that makes these turf demands sound so loud. (p. 134)

While the lawn is understood in a positivistic light as being an object disciplined by the knowledge produced through science and technology, the lawn actually operates as a de facto agent, instructing and disciplining that same social mind.

This positivistic distortion arises, at least partially, from the nature of human knowing. When we come to perceive, understand, feel, and imagine the world from within one complete set of mediational means, other perspectives naturally become less accessible and imaginable (Ryan, 2015). The picture of wild bee balm on the frontispiece (see Figure P.1) illustrates this phenomenon quite well. The camera's focus on the bee balm puts everything in the background out of focus. It takes some effort to consider anything from more than one perspective. In fact, from some research being done in transformational learning, it seems that often the jump from one paradigm to another is either simply not achievable at all, or is only sustainable for short periods because the rest of the entire constellation always tends to pull the mind back to the original default setting (Tschakert et al., 2016). The current paradigm has a natural advantage in the sense that some path dependency has already been established, both in the synaptic and muscle structures of body-brains and also as material artifacts, like lawns, that acquire some ability to act on their own. This problem is often rooted in a failure to take materiality seriously, an implicit assumption that mind operates upon material that is external to and separated from mind, but, of course, knowledge rarely counts for much until it is materialized and enacted into shared practices using physical parts of the world. The Western-Cartesian perspective amounts to a denial that we even could hear the lawn's demands for water and pesticides, which Robbins (2007) claims are "so loud" (p. 134). Yet positivism additionally both explicitly and systematically suppresses other competing ways of knowing, other competing ways of making and materializing, and other ways of becoming other voices or traces of forgotten voices making different demands.

Regardless of whether positivism might be thought of as a Western paradigm rather than an entire culture, there can be no denying that Western knowledge-production processes have generally suppressed other viable ways of producing actionable knowledges. But if, as already argued in the opening pages, knowledges should be evaluated by what they do in the world, then by ignoring the performative aspects of knowledge production, Western knowledge systems paradoxically cede any hope of steering society toward a hoped-for future. Instead, the agency directing the unfolding of human development is surrendered to the methods, tools, and instruments of Western knowledge production.

This might be contrasted with Newhouse's (2004) proposed paradigm of "complex understanding" (p.143) which shows awareness of the implicit Western knowledge trap and seeks to place other forms of knowledge production on an equal footing:

> Complex understanding occurs when we begin to see a phenomenon from various perspectives, as well as the relationships among these perspectives. Complex understanding does not seek to replace one view with another but to find a means of ensuring that all views are given due consideration. It does not work in an either-or fashion. A phenomenon is not one thing or another, but all things at one time. Complex understanding allows for our understanding to change, depending upon where we stand to see or upon the time that we look or who is doing the looking. Complex understanding is grounded in a view of a constantly changing reality that is capable of transformation at any time. (p. 143)

Normally, though, Western scientific knowledges *will* act to delegitimize other knowledges, *will* deny the inevitability and the potential benefits of ignorance, and *will* conceptualize those other knowledges as being *in competition*. Positivism *will* confer a distorted perception, understanding, and imaginary through such integral components as, for example, the denial of Indigenous knowledges (Newhouse, 2004) or tacit knowing (Polanyi, 1966, pp. 1–25) or by the reduction of objects of study to abstract essences. For example, it is routinely considered legitimate to transfer knowledges produced in laboratory settings to broader, different, and much more complex settings. Technological innovations continue to be misperceived as progress, and reflexivity about the ways the knowledges we design continue designing once released into the world is also generally overlooked. Positivism *will* hold and enforce the tenet that knowledge should never be anything but universal—that there is always only one best and true answer to every question. Finally, Western knowledges *will*, for the most part, ignore the way matter matters and *will* deny the fluidity of connectivity between matter and mind. This has huge implications for sustainability, though there may be signs that this is beginning to change.

According to Popa, Guilermin, and Dedeuwaerdere (2015),

> The last decades have witnessed a growing interest in sustainability research. Global environmental change (including climate change, loss of biodiversity or deforestation), population growth, rising inequalities and the financial crisis are requiring policy action backed up by reliable scientific data. However, the mainstream scientific methodologies are often poorly equipped to deal with complex sustainability problems. In particular, solving sustainability problems involves decisions on values that require civic participation and the building of social legitimacy for proposed transition pathways to sustainable societies. Therefore, both scientists and policy makers have called for re-conceptualizing the role of experts, practitioners and citizens in the production and use of scientific knowledge. (p. 45)

The legacy of positivism prevents us from understanding, much less being able to act on, sustainability. Positivism is part of a larger "ecology of bad ideas" (Bateson, 1972/2000) that "branches out like a rooted parasite through the tissues of life, and everything gets into a rather peculiar mess" (p. 492). Positivism molds society's perception and understanding of science, nature, humankind, fate, and the ethical, at a rational, but also a visceral, affective, *and material* level. Positivism, the cornerstone of the shared Western myth system, acts as a dominant, invasive, and keystone species in the ecology of bad ideas forming our contemporary unsustainability.

The irony, or paradox, of Western knowledge is that it systematically produces exactly what it wants to avoid: regimes of fire suppression result ultimately in increasingly terrible forest infernos, antibiotics in superbugs, Maximum Sustainable Yields with resource collapse, etc. As Bateson and Bateson (1987) quite rightly point out, "we become that which we pretend" (p. 182) and this applies as much to knowledge/meaning systems and models of subjectivity as it does to societies of peoples and to individual person-tool-environment assemblages. The diseases brought from Europe to Turtle Island were not a sign of European superiority, but rather part of a long history of ecological incompetence that continues unabated, positivist delusions notwithstanding. The resilience and productivity of any system with ecological characteristics lies partly in the diversity of its species, the nature of relations between those species, and the resonance, tuning, or correspondence of those communities of species with environment, where environment is defined as that which constitutes and is constituted by the conscious, subconscious, organic, and inorganic actions and structures of a place or region. These are the relevant parameters to look at when examining the role of positivism in our neurotic social mind, for positivism relies on reductionism and atomism as much as Monarch butterflies rely on milkweeds.

Capitalism

The climax of this "ecology of bad ideas" (Bateson, 1972/2000, p. 492) is capitalism. The market, in the conceptual universe of capitalism, acts as the foolproof wind used to separate wheat from chaff. The market puts things where they belong and rewards those who merit rewarding. But as McGowan (2016) rightly observes, "Though capitalism demands that subjects act out of their self-interest, it sustains itself through their self-sabotage" (p. 73). If the gratification promised through consumerist logic were ever successfully achieved, capitalism would falter from the resultant collapse of aggregate demand. There is ultimately, then, nowhere left for capitalism to go as its operation leads to neither satisfaction nor sustainability. The logic of capitalism exists only as the consequence of, and in concert with, many other bad ideas (cultural tools, mediational means) and dysfunctional materializations of those same bad ideas.

These unsustainability cultural tools beckon and coax the physical environment into increasingly psychotic combinations and reconfigurations that then loop around to provoke even worse cultural tools to impair more minds. Unlike Odysseus, we lack any crew who could lash us to a mast and thereby save us from these siren calls and ourselves. There is no safe ground, no place to stand when even Nature, as a force independent of humans, has been pronounced dead (McKibben, 1989), and is now mainly memorialized and trivialized through ritualized recycling programs or public incantations dutifully repeated every April 22.

We might refer to this situation as a form of Western-Cartesian psychosis, where *psychosis* is used in much the same way Rieber and Green (1989, pp. 48–89) (borrowing from Freud, 1901 Translated by Brill 1914) discuss "the psychopathy of everyday life." They state:

> There has been a general trend toward a greater degree of normalized, or acceptable, antisocial behaviour in our society generally—the psychopathy of everyday life—and that this trend in turn can be meaningfully related to an increase of what will be termed "social distress." (p. 49)

Social distress, in turn, can be linked directly to unsustainability (Gifford & Gifford, 2016) as opposed to the connectivity characteristic of healthy minds and patterns of psychological development (Graham, 2014). Including material culture and cultural landscapes within Rieber and Green's analysis of psychopathy makes it much easier to understand how social distress is distributed in the same way cognition and meaning are always extended—through brain-body-tool-environment assemblages. McGowan (2016) provides a good example of how psychoses escape from brains, dribbling away into the confines of homes and gardens, factories and fields:

> The absence of public space is not simply a problem for the lower classes that cannot afford entry into amusement parks where their children can play or gated neighbourhoods where they take a stroll without worrying about violence. The privatization of the commons also represents a retreat from subjectivity itself and from the way the subject satisfies itself. The subject is inherently a public being: its subjectivity forms through its interaction with the desire of the Other. (p. 52)

The "absence of public space" emerges from the mastering logic of the market forming a default setting for a shared model of (capitalist) subjectivity. Systems of ethics designed with these capitalist tools, such as utilitarian or Kantian models, also exclude a public space and subjectivity constituted through recognizing the desires of Others. Utilitarianism, like mainstream economic theory, instantiates an atomistic subject-object type of relation.

Likewise, Kantian approaches aspire to a position of disinterestedness that, quite frankly, represents a disassociation from reality (Berleant, 1994). They operate upon Others conceptualized-as-separate, where the separateness is not limited to geographical space, but extends into fields of concern, empathy, and caring. Of course, from the perspective of the extended and distributed mind, the Other (including McGowan's more limited sense of the *human* economic actor "Other") is technically not separate at all, nor limited to humans, but includes other impinging forms and actions of material artifacts and cultural landscapes. From other perspectives, there is no separateness. But the absence of public space for children's play is just one expression of the materialization of subject-object dualism mediated through capitalism's disinterested social mind. This does not, then, represent an ethical failure of the individual to adequately provision children with the necessities of a healthy becoming. The ethical failure is rather distributed among people, places, and cultural tools. It emerges from the quest to enact a technically impossible disinterestedness and maximization of utility as normalized within the economic institutions of *real estate development*.

How could we expect to build any sustainable ethical framework if our tools and materials are unwittingly designed and inherited and consequent upon this form of Self-destruction? (Self is capitalized here as shorthand to denote the brain-body-tools-environment assemblage as opposed to the more vernacular understanding of a being contained entirely within a membrane of skin.) Taking children away from parents, family, and land to be "educated" in the Canadian residential-school system, for example, could be considered as just such a form of Self-destruction. The Self destroyed here in the project of "taking the Indian out of the child" (Popeski, 2016, n.p.) is not limited to the Indigenous child, or even entire Indigenous cultures; the person-tool-environment assemblage even works to trigger defensive (yet never wholly successful) cognitive tools in the kidnappers and re-educators. If, as Hundert (1997) argues, we can know murder is ethically wrong because it is so psychologically damaging to the perpetrator, we might also imagine psychological foreclosures inherent in the "job" of government-sanctioned child snatcher or child brainwasher. Once we move away from the impoverished perspective of Cartesian dualism, we can see how bad ideas become materialized into broader bad ecologies. Yet the destructive impact spreads throughout the extended mind, including the body-brain, mediational means, surrounding lands, and material artifacts, including victims *as well as* perpetrators. The consequences of capitalist mediational means are as obvious in the contemporary destruction of rain forests as in the very intentional cultural genocide of the residential-school system. Yet the atomistic cultural tools of positivism and capitalism continue to provide cover for a system that fails to adequately provide the fundamental requirements for the healthy human socio-psychological development of children, whether in the creation of public spaces; provision of physical, affectual, and loving care; or even security from not being snatched by government employees.

McGowan's point (echoed by Beatley, 2005; Benedict, 1989; Clark, 2002; Fry, 2012; Hundert, 1997; Louv, 2005, 2011; and many others) is that when capitalist logic exercises a determining effect on the configuration of the (anti-)social mind, the result is at odds with the most basic requirements for socio-psychological health in humans. The world becomes fractured and sorted into pieces according to their value to the ghost in a machine. With only pieces like these to work with, reanimating and rejoining the world, (re-conceptualizing the way "matter matters")—recognizing and accepting the many forms of agency that the material world exercises—is much more than a problem of understanding. Yet only this alteration of mediational means will reveal the arbitrary nature of those superstitions of utility and disinterestedness that might then fall away as belief in Santa Claus eventually does for every child.

If people, mediational means, and environment interact together in unbroken loops, the problem of sustainability can be re-formulated as the problem of how to reintegrate ourselves into the loops in a way that changes the entire constellation. As already argued in the section on Positivism, the solution cannot be just a matter of epistemology. However, recognizing how knowledge, behavior, self, and situation are often *tacitly* embodied may still make a good place to start, as Polanyi's (1966) discussion of tacit knowing suggests:

> The declared aim of modern science is to establish a strictly detached, objective knowledge. Any falling short of this ideal is accepted only as a temporary imperfection, which we must aim at eliminating. But suppose that tacit thought forms an indispensable part of all knowledge, then the ideal of eliminating all personal elements of knowledge would, in effect, aim at the destruction of all knowledge. The ideal of exact science would turn out to be fundamentally misleading and possibly a source of devastating fallacies. (p. 20)

Tacit knowing is directly related to life's predisposition to come into correspondence with a material context in terms of isomorphy, tuning, and resonance (Knappett, 2005, p. 61). But, as the case of capitalism reveals, that coming into correspondence can include the mastery of hallucinatory cultural tools, so that, for example, we mistake marriage partners for commodities (Becker, 1973), or land for real estate, or think money more vital than food. Sustainability then becomes initially a problem of dissociation from those internalized (mastered) aspects of acquired (sociocultural) identity and their replacement with other less hallucinatory tools (Graham, 2014).

Situated (Un-) Sustainability

Any dissociative shift into a new paradigm would, as Polanyi's emphasis on tacit knowledge suggests, entail a shift in how we think about thinking. We tend to think of thought as something that occurs individually inside the brain of one person. Yet our perception and thought is clearly always embodied, situated, extended, and sociocultural; it is epistemic *and* ontological. Bateson (1972/2000) refers to this hallucinatory split between epistemology and ontology as "awkward" and points out that "it is … incorrect to suggest that they are separable in human natural history" (p. 314). Cognitive processes require bodies, including all the muscular, skeletal, endocrine, neural, bacteriological, and other myriad systems that being embodied entails. They also require a physical place. Sensory deprivation, especially for extended periods, for example, impedes the normal development of perceptual and cognitive functions (see also Hundert, 1997, pp. 115–128):

> Despite the many examples of such complex cycles of interaction now known, body and brain are usually conceptualized as separate, in structure and function. The idea that it is the entire organism rather than body alone or brain alone that interacts with the environment often is discounted, if it is even considered. Yet when we see, or hear, or touch or taste or smell, body proper *and* brain participate in the interaction with the environment. (Damasio, 1994, p. 224)

This comingling of person-environment also alters the brain and the body; all the pieces of the perceptual and cognitive process are *moving* pieces. Life is always a *dynamic* and *situated* affair.

Life supplies us with a suite of cultural/cognitive tools to extend our minds into the world. These are acquired just by being among other peoples, and we put those available tools to use, firing and activating specific synaptic connections and leaving other potential connections to wither on the vine. These tools can be structured, for example, in the form of languages (Bateson, 1989).

Learning a language really entails the acquisition of a mind by that language; a filter through which the world will then become known. The same is true of any material artifact representative of any specific material culture, whether they be stone hand axes, hammers, fishing poles, or video lottery terminals. Identity, both individual and societal, emerges from the available brains and bodies engaged in the psycho-cultural communion with tools in places over time. This continuous marriage of structure and flow constitutes an ecology in which mind and matter dissolve into each other, forming the outlines of human experience:

> People create time and space through their actions. Time and space, in turn, become part of habitual action, shaping the nature of reference between actions. Time and space are created on a number of levels through social action. (Gosden, 1994, p. 34)

How, then, to understand the epistemic-ontological ground that might eventually grow a new sustainable person-tool-environment mind and a corresponding ethics?

Gosden (1994) has beautifully described that territory and its awesome potential: "The bulk of life has a particular structure of recursiveness, drawing on the resources contained in human bodies and material things. The mass of unthought practice gives life a general forward movement and direction" (p. 125). At perhaps its simplest level, we could think of the bulk of life as evidenced in the child-bobo doll dyad of Bandura's experiments (Bandura, 1963; Graham & Arshad-Ayaz, 2016). Bandura arranged for children to witness adults putting inflatable plastic bobo dolls into a practice of aggression by hitting, punching, kicking, hammering, and throwing into the air the clown-faced doll. The children, when left alone with the doll, would generally imitate most, if not all, of the same repertoire of aggression. That repertoire, in other words, was the product of a type of implicit learning, mastered as a tacit knowledge that could only be partially and quite imperfectly communicated in words. That assimilated repertoire would then form a small part of the bulk of life, but it would also include the programming of violence on television, the notable violence in American society, the formative years of Bandura, the formulation of the research question, the experiment-design process taking place within a specific historical-cultural context, and so on. The ecology of cultural tools, both material and ideational, is not that different from an ecology of flora and fauna whose character emerges from such factors as a bedrock of either limestone or granite, precipitation and temperature regimes, biodiversity, nutrient cycling, and so on. All these various matters matter.

Now, generally speaking, ethics only addresses part of this bulk of life. While there have certainly been many productive, necessary discussions of ethics consistent with a hypothetical sustainability-tooled social mind (see, for example, Evernden, 1985; Naess, 1990; Plumwood, 2002; Smith, 2011; Stefanovic, 2000), from my perspective, these efforts still represent the cart and not the horse. The horse, the engine driving social change, is Gosden's "bulk of life." In other words, ethics need to be extended, distributed, and situated throughout environment rather than focused on simply changing thoughts, values, and ideas. Education for sustainable development confronts the same problem: students face a breach between the various knowledges of sustainable development and the already-lived-in world of Self-destruction. A description of an ethics that does not fit or mesh with the already-lived-in world is like a tool that nobody quite knows how to use. In the end, the incongruence between tool and world generally reinforces the dominant shared ideas about the way the world works, including a delusional focus on propositional knowledges, while simultaneously obscuring and delegitimizing the more important narrative knowledges that guide the various processes of human development (see Tsoukas, 2005, pp. 69–93). So much of ethics and education is focused on the facts—the growing list of provable, demonstrable, and consistent, carved in stone, known knowns. Going on below the radar, however, is a mostly subconscious story about the matter that matters to us and the matter that simply does not. That mattering is prescribed by the world as we encounter it as much as it is by the knowledges, practices, and stories that also pre-exist us and that we come to be acquired by. These currently operate in tandem and continuously as a veritable dynamo of unsustainability. This is a critical, yet little explored, part of the domains of (un-) sustainability.

Material Engagement Theory (Malafouris, 2013a, b), which I turn to in the next chapter, provides the primary tool I will use to get at Gosden's "bulk of life." Emerging from the academic field of cognitive archaeology, MET has been used to put forward a theory of the emergence and development of the human mind, whose cerebral component has changed little from its origins and is still recognizable in the characteristics of its operation today. It is a mind that is activated within and co-constituted through the engagement process of tool use, both material tools and cognitive-cultural tools that give structure to perception, cognition, and emotion. My primary focus will also be on the social and material rather than on some imagined fantasy of a context-free individual with a monopoly on the exercise of agency. This MET perspective emerges from, builds upon and complements the researches of various scholars working across many academic disciplines, including sociologist and cognition scholar Zerubavel (1997) who reminds us, "we *learn* how to focus our attention, frame our experience, generalize and reason in a socially appropriate manner" (p. 13); rhetoricians (Barnett & Boyle, 2016; Wallace, 2016) who consider the many ways matter *speaks*; and material-culture theorists who develop new ways of understanding how things become social and have life histories (Appadurai, 1986/2010).

This book continually draws parallels between the individual mind and the social mind. The individual mind is minimally constituted from three components: first, a brain putting into practice a set of sociocultural tools that together form a culturally guided model of subjectivity; next, a body connecting brain and world in unbroken circuits of experience; and, finally, a place providing the available tools and all the material stuff with which to put those tools, bodies, and brains into practice, as well as the field for operating in. The social mind is composed of many individual minds woven into the same circuits using the same, or very similar, tools in perceived-as-operationally-similar places. Society, like an individual mind, brings certain objects and events into focus by giving less attention to other objects and events. Like the emergence of a model of subjectivity, (or personality, if you like) from the firing of certain brain synapses and the atrophy of others, the social mind emerges from the available sociocultural tools within environment (composed of various brains, bodies, sociocultural frameworks, material artifacts, and physical places) to form the woven pattern of cultural becoming.

Damasio (1994) describes the problems with the dominant Cartesian model forming the contemporary social mind as follows:

> Versions of Descartes' error obscure the roots of the human mind in a biologically complex but fragile, finite, and unique organism: they obscure the tragedy implicit in the knowledge of that fragility, finiteness and uniqueness. And where humans fail to see the inherent tragedy of conscious existence, they feel far less called upon to do something to minimize it, and may have less respect for the value of life. (p. 251)

My reading of "where humans fail" can be translated as—where human fall under the spell of a psychotic unsustainability social mind, material culture, disfigured landscape, and corresponding set of cultural/cognitive/affective tools. To reiterate: This failure is not, in other words, just an ethical failure of an individual. There was, after all, something about Canada and Canadian society that enabled the Government employment posting for "child snatcher," and the hiring, training, and paying of people to kidnap children from their homes. Tools, both cultural-cognitive and material, were designed, made, and pressed into service to further the enterprise of kidnapping children as part of a cultural genocide that lasted over several generations. Like unsustainability, it was an enterprise that enlisted complicity from all quarters, human, nonhuman, living, and inert. This enterprise cannot, in fact, be separated from the systematic destruction and impoverishment of the land carried on since the invasion of Turtle Island by the first Europeans. This was not simply an ethical mistake traceable to some individual who proposed a policy or someone else who carried out that policy; it was a holistic ethical collapse and societal descent into collective ecological insanity coursing through minds, bodies, tools, and entire environments.

The challenge of displacing unsustainability must also be understood as a challenge of diagnosing and treating this type of collective moral failure extending throughout our inherited, already-lived-in world. Remediation efforts need to be designed for that scale and that locus of pathology. Sustainability is very much a moral issue, but the concept of ethics needed to address our contemporary unsustainability needs to be vastly expanded to encompass situated and extended minds operating within ecologies of distributed agencies.

This exploration of the (un-) sustainability mind will often take the form of a serpentine wandering rather than the more linear style of argumentation. The map of sustainability must necessarily be speculative, guided by traces that are often obscure and difficult to uncover, recover, or discover, because the things we search out are always mediated things. There is never *just* a thing, there is never *just* a relationship, and there is never an unmediated anything. Any *thing* that we can come into relationship with necessarily comes through our mediational means. Insofar as we cannot escape our inherited and delusional mediational means, we are ill equipped and challenged to seek and achieve sustainability. Perhaps, though, we can still identify practices that might constitute an image of a good mind that might achieve a healthy and Self-sustaining balance based on a distributed and partnering agency between brain, body, tools, and environment.

A Map of the Book

The next chapter, "The Materially Engaged (Un-) Sustainability Mind," expands the general comments of Chapter 1 about mediated action and

MET into a theoretical perspective that informs all following discussions of (un-) sustainability. That theory of (un-) sustainability, as revealed through a far-from-exhaustive set of variables (bodies, tools, places, processes, agencies, memories, and atmosphere) will then be employed in a series of contexts to illustrate particular problems and possible diagnoses. This is first demonstrated with a simple but concrete example in Chapter 3, "Let's go Fishing." This preliminary description of a simple gesture, a parent inviting a child to go fishing, provides an important micro-level illustration of the structures, processes, and countless threads of action and artifact that weave such a seemingly simple gesture into ways of becoming less or more sustainable.

Chapters 4 and 5 explore the background aspect of the Cartesian inheritance. Chapter 4, "Economies of Mind," maps out the contours of Polanyi's (1977) prerequisite for a market economy: the "marketing mind" (p. 1). Like ethics, economic systems reflect the epistemological and ontological grounds they sprout from and grow in, the stories used to tell other stories, and the matter that matters. The goal is to help economics remember history (Hodgson, 2001) and to focus especially on the *mind* of an economic system. Just as Malafouris (2013a) "perceive(s) early stone tools as capable of transforming and extending the cognitive architecture of our hominin ancestors" (p. 233), this chapter identifies the material tools within a constellation of practices leading to the emergence of the economistic mindscape.

Chapter 5, "Materialities of Mundane Terror," builds on Väliaho's (2014) discussion of those treatments of PTSD that seek to make the initial causative experience mundane by repetitively re-experiencing the trauma through virtual reality. This treatment regime is directed at the "affective and emotional charges related to traumatic events, rather than conscious memories of the events, that are significant" (p. 71). Its goal is to

> evoke and animate the bodily memory, and because they do so repeatedly, they are meant to trigger "habituation" and the extinction of the fear response—the notion of habituation deriving from Pavlovian conditioning, which interiorizes the source of affects within the organism's nervous system and conditioned actions. (p. 71)

I compare this deliberately designed process to the inherited and normalized terror of everyday life, our habituation to trauma, and to the very mundane futurelessness that now form the background for socio-psychological development.

The following chapter, "Mowed (Un-) Sustainability Minds," traces the unthought model of capitalist subjectivity into the realm of gardens and the various philosophies and human-place relations that gardens enable, direct, embody, enliven, and conjure into becoming. The correspondence between a macro level metanarrative and the subconscious and ritualistic practice of lawn keeping provides the background for a more hopeful theory of transformational change that includes the way matter matters.

Chapter 7, "Tracing the Absences of Sustainability," is a personal reflection on the kinds of agency absences can exercise, the potential positive role absences could play for moving society onto a path of sustainability, and some thoughts on the necessarily ambiguous path before us. Finally, Chapter 8 provides a summary of the main arguments and points the way to future research and a shift out of the Western-Cartesian unsustainability trap. I propose a few possible "steps" in the direction of a materially engaged sustainability ecology of mind. These steps always need to consider local context but also must involve more than ideas because changing our fundamental relationship with the world in the direction of sustainability absolutely involves changing the whole world.

References

Appadurai, A. (2010). *The social life of things: Commodities in cultural perspective*. Cambridge, UK: Cambridge University Press. (Original work published 1986).

Bandura, A., & Walters, R. H. (1963). *Social learning and personality development*. New York, NY: Holt, Rinehart, and Winston.

Barnett, S., & Boyle, C. (2016). *Rhetoric through everyday things*. Tuscaloosa: University of Alabama Press.

Barry, J. (2012). *The politics of actually existing unsustainability: Human flourishing in climate-changed, carbon-constrained world*. Oxford, UK: Oxford University Press.

Bartolo, K. (1998). Bathing babies: Empirical segmentation and theoretical contextualization. *Psychological inquiry*, 9(4), 274–276.

Bateson, G. ([1973] 2000). *Steps to an ecology of mind*. Chicago, IL: University of Chicago Press. (Original work published 1972).

Bateson, G., & Bateson, M. C. (1987). *Angels fear: Towards an epistemology of the Sacred*. New York, NY: MacMillan.

Bateson, M.C. (1989). Language, languages, and song: The experience of systems. In R. W. Rieber (Ed.), *The individual, communication, and society: Essays in memory of Gregory Bateson*. Cambridge, UK: Cambridge University Press.

Beatley, T. (2005). *Native to nowhere: Sustaining home and community in a global age*. Washington, DC: Island Press.

Becker, G. S. (1973). A theory of marriage: Part I. *Journal of Political Economy*, *81*, 813–846.

Benedict, R. (1989). *Patterns of culture*. Boston, MA: Houghton Mifflin. (Original work published 1934).

Berleant, A. (1994). Beyond disinterestedness. *British Journal of Aesthetics*, *34*, 242–255.

Boivin, N. (2008). *Material culture, material minds: The impact of things on human thought, society, and evolution*. New York, NY: Cambridge University Press.

Boulding, K. E. (1978). *Ecodynamics: A new theory of societal evolution*. Beverley Hills, CA: SAGE.

Bruner, J. (1986). *Actual minds, possible worlds*. Cambridge, MA: Harvard University Press.

Burke, K. (1984). *Permanence and change: An anatomy of purpose* (3rd ed.). Berkeley: University of California Press.

Capra, F. (1996). *The web of life: A new understanding of living systems*. New York, NY: Anchor Books.

Capra, F. (2010). *The Tao of physics: An exploration of the parallels between modern physics and Eastern mysticism* (5th ed.). Boston, MA: Shambhala. (Original work published 1975).

Carter, J. A., & Palermos, S. O. (2016). Is having your computer compromised a personal assault? *Journal of the American Philosophical Association*, *2*, 542–560.

Clark, A., & Chalmers, D. (1998). The extended mind. *Analysis*, *58*, 7–19.

Clark, M. C. (2002). *In search of human nature*. New York, NY: Routledge.

Cole, M. and Scribner, S. (1978). Introduction. In L. V. Vygotsky (Ed.), *Mind in society: The development of higher psychological processes*. Cambridge, MA: Harvard University Press.

Cosgrove, D. (2001). *Apollo's eye: A cartographic genealogy of the Earth in the Western imagination*. Baltimore, MD: John Hopkins Press.

Cronon, W. (1983). *Changes in the land: Indians, colonists and the ecology of New England*. New York, NY: Hill & Wang.

Crutzen, P. J., & Stoermer, E. F. (2000). The "Anthropocene." *Global Change Newsletter*, *41*, 17–18.

Dale, A. (2001). *At the edge: Sustainable development in the 21st century*. Vancouver: University of British Columbia Press.

Damasio, A. (1994). *Descartes' error: Emotion, reason, and the human brain*. New York, NY: Penguin Books.

Davidson, J., & Smith, M. (2003). Bio-phobias/techno-philias: Virtual reality exposure as treatment for phobias of "nature." *Sociology of Health & Illness*, *25*, 644–661.

Diamond, J. (1999). *Guns, germs, and steel: The fates of human societies*. New York, NY: Norton.

Enz, J. J. (1976). Origin of the dualism expressed by "sons of light" and "sons of darkness." *Biblical Research*, *21*, 15–18.

Evernden, N. (1985). *The natural alien*. Toronto, ON: University of Toronto Press.

Fernandes Mendes, D. L., & Seidl-de-Moura, M. L. (2015). Facial expressions in mother-baby interactional contexts and emotional development. *Journal of Communications Research*, *7*, 373–389.

Fisher, A. (2013). *Radical ecopsychology: Psychology in the service of life*. Albany, NY: SUNY Press.

Forgas, J. P., Williams, K. D., & Wheeler, L. (Eds.). (2001). *The social mind: Cognitive and motivational aspects of interpersonal behaviour*. Cambridge, UK: Cambridge University Press.

Franklin, U. (1991). The Real World of Technology. Toronto, ON: Anansi.

Freud, S., & Brill, A. A. (1914). *Psychopathology of everyday life*. New York, NY: Macmillan.

Fry, T. (2012). *Becoming human by design*. London, UK: Berg.

Gell. A. (1998). *Art and agency: An anthropological theory*. Oxford, UK: Clarendon Press.

Gifford, E., &Gifford, R. (2016). The largely unacknowledged impact of climate change on mental health. *Bulletin of the Atomic Scientists, 72*, 292–297.

Gosden, C. (1994). *Social being and time*. Oxford, UK: Blackwell.

Graham, P. (2014). *Meaning, education, & sustainability: Building connectivity through dissociation* (Unpublished Master's thesis). Concordia University, Montreal, QC.

Graham, P., & Arshad-Ayaz, A. (2016). Learned (un-) sustainability: Bandura's bobo doll revisited. *Journal of Education for Sustainable Development, 10*, 262–273.

Haraway, D. (2016). *Staying with the trouble: Making kin in the Chthulucen*. Durham, NC: Duke University Press.

Harrison, R. P. (1992). *Forests: The shadow of civilization*. Chicago, IL: University of Chicago Press.

Heidegger, M. (1962). *Being and Time*. London, UK: SCM Press.

Hodgson, G. M. (2001). *How economics forgot history: The problem of historical specificity in social science*. London, UK: Routledge.

Hoffmeyer, J. (2009). *Biosemiotics: An examination into the signs of life and the life of signs*. Scranton, PA: University of Scranton Press.

Hundert, E. M. (1997). *Lessons from an optical illusion: On nature and nurture, knowledge and values*. Cambridge, MA: Harvard University Press.

Ingold, T. (2000). *The perception of the environment: Essays in livelihood, dwelling and skill*. London, UK: Routledge.

Ingold, T. (2013). *Making: Anthropology, archaeology, art and architecture*. London, UK: Routledge.

Kim, S., & Kochanska, G. (2017). Relational antecedents and social implications of the emotion of empathy: Evidence from three studies. *Emotion, 17*, 981–992.

Knappett, C. (2005). *Thinking through material culture: An interdisciplinary perspective*. Philadelphia, PA: University of Pennsylvania Press.

Lakoff, G., & Johnson, M. (1980). *Philosophy in the flesh: The embodied mind and its challenge to Western thought*. New York, NY: Basic Books.

Latour, B. (2005). *Reassembling the social: An introduction to actor-network-theory*. Oxford, UK: Oxford University Press.

Lave, J., & Wenger, E. (1991). *Situated learning: Legitimate peripheral participation*. Cambridge, UK: Cambridge University Press.

Lear, J. (2006). *Radical hope: Ethics in the face of cultural devastation*. Cambridge, MA: Harvard University Press.

Lopes, A. M. (2017). Ontological design as an ecological practice. *Arena, 47/48*, 172–191.

Louv, R. (2005). *Last child in woods: Saving our children from nature-deficit disorder*. Chapel Hill, NC: Algonquin Books.

Louv, R. (2011). *The nature principle: Reconnecting with life in a virtual age*. Chapel Hill, NC: Algonquin Books.

Lovelock, J. (1989). *Gaia: A new look at life on earth*, Oxford, UK and New York, NY: Oxford University Press.

Lovelock, J. (1991). *Healing Gaia: Practical medicine for the planet*. New York, NY: Harmony Books.
Lund, K. A., & Benediktson, K. (2010). *Conversations with landscape*. Farnham, Surrey, UK: Routledge.
Macy, J., & Brown, M. (1998). *Coming back to life: Practices to reconnect our lives, our world*. Gabriola Island, BC: New Society.
Malafouris, L. (2013a). *How things shape the mind: A theory of material engagement*. Cambridge, MA: MIT Press.
Malafouris, L. (2013b). Mind into matter. *New Scientist, 219*(2933), 28–29.
Malafouris, L. (2015). Metaplasticity and the primacy of material engagement. *Time and Mind*, 8(4), 351–371.
McGowan, T. (2016). *Capitalism and desire: The psychic cost of free markets*. New York, NY: Columbia University Press.
McKibben, B. (1989). *The end of nature*. New York, NY: Random House.
Mead, G. H. (1934). *Mind, self and society*. Chicago, IL: University of Chicago Press.
Mishra, P. (2017). Politics in the age resentment: The dark legacy of the Enlightenment. In H. Geiselberger (Ed.), *The great regression*. Cambridge, UK and Malden, MA.: Polity Press, 104–116.
Morgan, G. (1997). *Images of organization* (2nd ed.). Thousand Oaks, CA: SAGE.
Naess, A. (1990). *Ecology, community and lifestyle*. Cambridge, UK: Cambridge University Press.
Newhouse, D. (2004). Indigenous knowledge in a multicultural world. *Native Studies Review*. *15*(2), 139–154.
Overmann, K. A. (2017). Thinking materially: Cognition as extended and enacted. *Journal of Cognition and Culture*, *17*(3–4), 381–400.
Plumwood, V. (2002). *Environmental culture: The ecological crisis of reason*. London, UK: Routledge.
Polanyi, K. (1977). *The livelihood of man*. New York, NY: Academic Press.
Polanyi, M. (1966). *The tacit dimension*. New York, NY: Anchor Books.
Popa, F., Guilermin, M., & Dedeuwaerdere, T. (2015). A pragmatist approach to transdisciplinarity in sustainability research: From complex systems theory to reflexive science. *Futures*, 65, 45–56.
Popeski, R. Contributor. (2016, February 12). Canada's most racist city? A year later, Winnipeg takes stock. *Christian Science Monitor*, p. N.PAG.
Renfrew, C. (2009). *Prehistory: The making of the human mind*. New York, NY: Modern Library.
Renzi, D. T., Romberg, A. R., Bolger, D. J., & Newman, R. S. (2017). Two minds are better than one: Cooperative communication as a new framework for understanding infant language learning. *Translational Issues In Psychological Science*, 3(1), 19–33.
Richter, D. K. (2001). *Facing east from Indian country: A native history of early America*. Cambridge, MA: Harvard University Press.
Rieber, R. W., & Green, M. (1989). The psychopathy of everyday life: Antisocial behaviour and social distress. In R. W. Rieber (Ed.), *The individual, communication, and society: Essays in memory of Gregory Bateson*. Cambridge, UK: Cambridge University Press.
Robbins, P. (2007). *Lawn people: How grasses, weeds, and chemicals make us who we are*. Philadelphia, PA: Temple University Press.

Ryan, P. (2015). Positivism: Paradigm or culture? *Policy Studies*, 36, 417–433.

Sahlins, M. D. (1963). Poor man, rich man, big-man, chief: Political types in Melanesia and Polynesia. *Comparative Studies in Society and History*, 5, 285–303.

Sahlins, M. D. (1966). *Stone Age economics*. New York, NY: Aldine de Gruyter.

Séville, A. (2017). From "one right way" to "one ruinous way"? Discursive shifts in "there is no alternative." *European Political Science Review*, 9, (3), 449–470.

Shiner, L. E. (1972). Sacred space, profane space, human space. *Journal of the American Academy of Religion*, 40(4), 425–436.

Smith, M. (2011). *Against ecological sovereignty: Ethics, biopolitics, and saving the natural world.* Minneapolis: University of Minnesota Press.

Stefanovic, I. (2000). *Safeguarding our common future: Rethinking sustainable development.* Albany, NY: SUNY Press.

Todd, K. (2001). *Tinkering with Eden: A natural history of exotics in America.* New York, NY: Norton.

Tschakert, P., Das, P. J., Shrestha Pradha, N., Machado, M., Lamadrid, A., Buragohain, M., & Hazarika, M. A. (2016). Micropolitics in collective learning spaces for adaptive decision making. *Global Environmental Change*, 40, 182–194.

Tsoukas, H. (2005). *Complex knowledge: Studies in organizational epistemology.* New York, NY: Oxford University Press.

Väliaho, P. (2014). *Biopolitical screens: Image, power and the neoliberal brain.* Cambridge, MA: MIT Press.

Valsiner, J., & van der Veer, R. (2000). *The social mind: Construction of an idea.* Cambridge, UK: Cambridge University Press.

Varey, W. (2011). Viability of psychological panarchy: Thought as an ecology. *Systems Research and Behavioral Science*, 28, 509–525.

Vygotsky, L. V. (1978). *Mind in society: The development of higher psychological processes.* Cambridge, MA: Harvard University Press.

Wallace, M. (2016). *Risk criticism: Precautionary reading in an age of uncertainty.* Ann Arbor: University of Michigan Press.

Weiss, A. S. (1995). *Mirrors of infinity: The French formal garden and 17th century metaphysics.* New York, NY: Princeton Architectural Press.

Wertsch, J. V. (1991). *Voices of the mind: A sociocultural approach to mediated action.* Cambridge, MA: Harvard University Press.

Wertsch, J. V. (1998). *Mind as action.* New York, NY: Oxford University Press.

Winnicott, D. W. (1971). *Deprivation and delinquency.* London, UK and New York, NY: Tavistock Publications.

Winston, M. L. (1997). *Nature wars: People vs. pests.* Cambridge, MA: Harvard University Press.

Zerubavel, E. (1997). *Social mindscapes: An invitation to cognitive sociology.* Cambridge, MA: Harvard University Press.

Zerubavel, E. (2015). *Hidden in plain sight: The social structure of irrelevance.* Oxford, UK: Oxford University Press.

· 2 ·

THE MATERIALLY ENGAGED (UN-) SUSTAINABILITY MIND

> Yet, despite the great flurry of activity around sustainability and the large number of actors now engaged in the sustainability mission, we still find ourselves in an increasingly dismal ecological position, one that holds out little promise for economic and social advancement. In fact, ecological systems are more fully understood now in the era of international global climate change research than they ever have been, but evidence indicates that our systems are more, not less, degraded each year. (Farley & Smith, 2014, pp. 2–3)

Farley and Smith's (2014) critique is characteristic of a broad-based discontent, within academia and across much of society, with the term *sustainability*. That discontent encompasses both the ambiguity of the term and the lack of measurable success in attaining it. Sustainability, simply put, seems to be unworkable given the current tools available to us for organizing and structuring our relations within environment. The blueprints and solutions imaginable from within the Western historical inheritance, even if plausible in theory, are often simply not feasible from a political standpoint, to say nothing of the social psychological obstacles they would encounter. There would seem to be simply no way out of this conundrum. And yet, sustainability has certainly been achieved by other societies in different times and places. It does not seem that the human organism is itself inherently unsustainable by nature. How then, might we account for and overcome these very considerable difficulties.

Mediated Action and (Un-) Sustainability

Perhaps concepts of mediated action and mediational means might bring some theoretical clarity to our understanding of the term *sustainability* and the reasons so many are so unhappy with the term (see, for example, Barry, 2012; Farley & Smith, 2014). Our necessary first step is to recognize that sustainability must be understood as a mediating cultural tool and not just an ideal. Like other mediating cultural tools, it cannot simply be applied and used in isolation but requires a broader context of other supporting and interlocking cultural tools. Sustainability needs the scaffolding provided by these other tools for it to become a meaningful concept associated with, enabling, and enabled by, ongoing sociocultural action. In that regard, it is critical to recognize that the dominant Cartesian mediational means, discussed above, provides little support for the term or the requisite institutions, artifacts, etc., needed to enact sustainability. It is a term describing a hoped-for, but still imaginary, process rather than any theoretically viable practice that would spark us to actively confront the overwhelming and frighteningly unavoidable evidence of self-inflicted harm. As an isolated term it can, at best, only disturb and provoke by pointing to the alarming normality of contemporary (unsustainable) life.

Although many critics, such as Farley and Smith, worry that sustainability has become overused and is consequently nearly meaningless, the problem is more about the way the tool is used and what it is used in conjunction with. People are perfectly capable of considering their shopping habits or their government's behavior from a sustainability perspective, but that is not the only tool they will use to conceptualize, feel, and cogitate about these things. The entire tool kit of mastered mediational means and how those tools are engaged within communities of practice will be the most important determinants of how the ecology of material engagement ultimately shakes out on the ground. Part of the problem of sustainability is that the cultural tool, sustainability, cannot be sustained within Western-Cartesian sociocultural action because other tools, such as profit maximization, resonate more profoundly than sustainability.

From within the Cartesian framework, sustainability also pushes up against a shared (social mind) conceptualization of the human capacity to know and apply knowledge to successfully manage the physical world. This Cartesian tool kit provides a false sense of ontological security through conceiving the world as a machine-like thing the parts of which can be examined and manipulated in isolation. However, this runs counter to any attempt to enact sustainability. Sweeny (2017) provides a good example:

> In the 1920s, local ranchers in the U.S. states of Montana, Wyoming, and Idaho called for the removal of wolves from Yellowstone National Park. Over the next seventy years, wildlife biologists and park rangers watched as the integrity or "wholeness" of the Yellowstone ecosystem unraveled. Without a top predator such as the wolf, the populations of deer and elk grew unchecked, which led to over-grazing of shrubs and plants, decreasing the nesting areas for migrating birds, changing the habitat for beaver colonies, and so on. (p. 143)

Here, an incomplete and isolated knowledge of the wolf as "livestock predator" resulted in far-reaching ecological effects that changed an entire landscape. The presumption that such isolated attention to one aspect of a problem is sufficient encourages exactly the kind of practices and actions that are impossible to reconcile with sustainability. Even more sophisticated models, such as systems theory or chaos theory, that recognize interconnectivity, are often still used in ways that maintain, and remain within, the supremacy of (human) knowledge and the dominance of thought over matter. From within the Western tradition, sustainability can only be understood as acting as something that involves measuring and weighing, maximizing and regulating, or, in short, knowing and controlling. Its focus naturally becomes the search for some type of formula to "solve" the problem of unsustainability. But the root of the problem of sustainability is not primarily a matter of the misapplication of Western-Cartesian mediational means but a feature of their normal operations.

Unfortunately, Cartesian mediational means seem to be very good not only at producing new "sciences" and new ways of framing knowledges but also at erasing awareness of the consequent ignorance that these same knowledge-production processes involve (Malewski & Jaramello, 2011). It becomes, in other words, practically impossible to evaluate inventions (e.g., the Hubble space telescope or the electron microscope or the knowledge spawned by those instruments) from any perspective but the one provided and enforced by our inherited cultural tools. As argued above, Western cultural tools are essentially epistemologically imperialistic, as they compete with and delegitimize other ways of knowing.

The only question that remains is whether we will perish with our mediational means intact or find new ways of becoming in the world with a new set of cultural tools co-constituting new sustainability persons. The practical difficulties associated with this proposition cannot be overstated because even changes previously thought of as revolutionary—changes such as the agricultural and industrial revolutions—actually represented the playing out of the existing mediational means, not their radical transformation. Sustainability must attempt to be the tool that puts mind and matter back together, overturning, rejecting, and replacing Cartesian tools with tools that would reflect and constitute a sociocultural practice of partnership between mind and matter.

This task is very far from easy, but we need to begin by recognizing that sustainability too needs to be considered in context. It is not accidental that sustainable cultures and communities generally do not even have terms to represent the concept of sustainability. The term *sustainability* represents an aspirational concept for the Western world, something that does not now actually exist (Barry, 2012) and something that does not "fit" the way the Western world has been conceptually and materially configured. So far, we have been mostly unable to manage, despite our best efforts, to enact and materialize sustainability on a scale that even slows, let alone halts or reverses, the state of ecological deterioration (whether this is measured by the acceleration of world-altering climate change, world-impoverishing biodiversity collapse, the explosive increase in toxic pollutants, etc.). But this is not necessarily because there is something wrong with the idea of sustainability per se. Rather, these important facets of unsustainability are the direct result of inherited patterns of ongoing socioculturally mediated action. It is in the processes and institutions, in the formation and maintenance of those patterns of dysfunction, that the key to understanding (un-) sustainability must be sought.

The Materiality of (Un-) Sustainability

One important place to look for traces of sustainability, to look for answers to the challenge of social change outlined in the previous section, is the neglected part of the extended and distributed mind that is matter, materiality, and physical environment. As already mentioned above (Chapter 1, pp. 35), Gosden's work (2005) provides an excellent starting point to begin the task of developing of a theory of the materiality of (un-) sustainability and to also set the stage for the introduction of the methodological tool of Material Engagement Theory (MET). As Gosden notes,

> The nature of social being for people will be structured by the education of their senses by the objects surrounding them in childhood, giving them a series of stances and presuppositions towards the world derived from local material culture. People crystallize out in the interstices between objects, taking up space allowed them by the object world, with our sense and emotions educated by the object world. (p. 197)

In mapping out the ground from which human (un-) sustainability springs, we need to look to what Gosden describes as "the interstices between objects" from which peoples are constituted or "crystallize out" (p. 197). Presumably persons might "crystalize out" somewhere on a spectrum, from sustainability persons to unsustainability persons. Ingold (2013), building on Heidegger, would argue, I think that what Gosden really means is the interstices between "things" and not "objects." Ingold makes a distinction between what merely exists (objects) and what is actively engaged with—that is, what continues and is drawn into and re-forms as assemblages and loci of human action (things). Objects simply exist, but things carry on, according to Ingold (p. 70). As someone with some experience as a home owner, I would tend to agree with Ingold that buildings, for example, are things and, as such, always remain never quite finished, although as any student of geology can attest, even rocks are never really "finished" except when they are perceived in human time. The home changes as the people in it change and the mental image of "home" evolves. The home holds meaning that is generally more tacit than explicit and especially as it "educates our senses" without our conscious awareness. We do not need to focus our attention on the home for it to continue to carry profound meaning and associations for us.

That said, even when buildings are experienced subliminally as objects, and even simply the unconscious experience of the carpentered corners, the "object" still manages to work its way into the synaptic structures of our brains, eyes, and our entire visual-perception systems, a point Hundert (1997, pp. 208–234) illustrates using the Müller-Lyer optical illusion. The Müller-Lyer illusion (see Figure 2.1) appears to people whose perception is attuned to buildings with corners, as in most modern Cartesian homes, as two lines of different length, when in fact the two lines are of the exact same length. Hundert (1997) explains the phenomenon as follows:

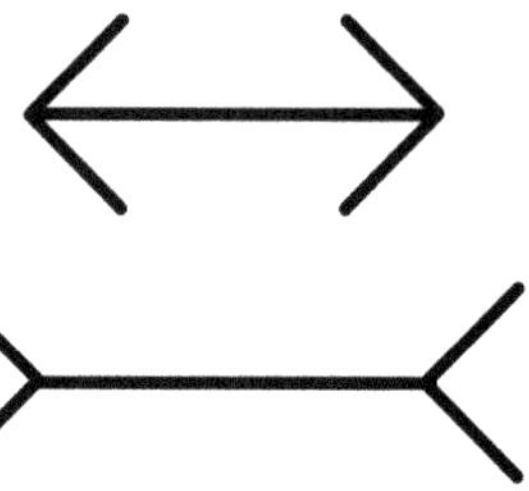

Figure 2.1 The Müller-Lyer Illusion. (adapted from Donaldson and Macpherson, 2017)

> In an environment filled with rectangular rooms, buildings, furniture, boxes, TV sets, and so on (the "natural" environment of most of us), right angles appear obtuse or acute, depending upon whether the corner is pointing away from us or toward us, respectively. This element of linear perspective was known to the ancient Greek and Roman artists in an intuitive way, but it was first "discovered" (intellectually) in the Italian Renaissance by Filippo Brunellesco (1377–1446), who was able to explain mathematically the curious distortions in so many classical works, such as the unrealistic appearing architectural background in Rome's famous Arch of Titus. Children raised in a rectilinear environment like ours spend their early years exploring those rectangular features of the world with their mouths and hands as well as their eyes, and their plastic visual input systems shape themselves to these realities using such cues to analyze depth in the visual field. Very early in life, our visual input systems have accommodated to an environment where obtuse angles are analyzed as 90° corners pointing away from us and acute angles as 90° corners pointing toward us. (p. 210)

Whether corners are consciously attended to or not, it does not seem that this process of adjusting one's visual perception to become susceptible to the trickery of the Müller-Lyer illusion involves any exercise of agency whatsoever. If anything, the corner is what exercises the agency over the wiring of visual perception systems. We do not consciously and deliberately educate our eyes and nervous system on how to interact with the world visually. In fact, the cause might even seem to be more logically assigned to the biosemiotic nature of the world we inhabit and that in-habits us.

The many ways that things and objects educate and form the operational patterns of our senses and sensibilities without any active engagement on our part is not limited to corners in rectangular buildings. To take a very different example, Kim et al. (2018) found that the health and diversity of the microbial ecologies in our guts have a direct correlation with our personality traits. Without any awareness on our part, we have also been "crystallized out" between the microbial matter living in, on, and around us. This is not to say that the microbes in the gut *cause* the personality traits or that the personality traits *cause* the ecology of gut biota, but only that it seemingly affects and/or is affected by who we are. Just as our perception attunes to the attributes of environment, so personality is attuned in response to aspects of environment of which we are rarely, if ever, consciously aware. People who live in high altitudes are acclimatized to air with less pressure and oxygen than people who reside in low altitudes, and so on. The world, like a parent, sets the ground rules, whether we like it or not. The distinction between things and objects, then, is like the distinction between my parent who teaches me how to tie my shoes and my parent who is in the next room keeping me from being frightened just by being there. Our brains, guts, and all the rest of us are "educated" by the world we inhabit, whether we are aware of it or not.

Bateson (2002) was also interested in this developmental process of sensory, aesthetic, and cognitive attuning, describes the apparent absence of agency, drawing parallels to numerous other forms of human-world tunings, as "a *zigzag ladder of dialectic between form and process*" (p. 182; italics in original).

> The very nature of perception follows this paradigm: that learning is to be modeled on the same sort of zigzag paradigm; that in the social world, the relation between love and marriage or education and status necessarily follow a similar paradigm; that in evolution, the relation between somatic and phylogenetic change and the relation between the random and the selected have this zigzag form. (p. 182)

Malafouris (2013a), who provides the main theoretical scaffolding for what has come to be known as Material Engagement Theory, reinforces this point somewhat more directly: "Cognition and action arise together, dialectically forming each other" (p. 74). Here, though we need to remember, as Malafouris too will emphasize, that first, perception, cognition, and emotion are not separable actions and second, following from the parity principle, (see Chapter 1, p. 23) materiality plays a constitutive role in these processes.

Our senses, thoughts, and emotions, in other words, unbeknownst to us, extend, seek out, adjust, and conform to the material world, including both things and objects. We do indeed "crystallize out" from the interstices between things/objects. That crystallization process requires the very specific aspects of brains, bodies, materiality, environment, and all these parts being configured into already-existing practices engaging mediational means. These sociocultural practices are not reducible to individual components but rather form one unitary system in action. Just as carpentered buildings form specific aspects of perception, lollipop-shaped ornamental shrubs and cities made for automobiles rather than people form specific aspects of our tacit knowledge of the way the world works whilst playing specific roles within that entirety.

It is therefore critical for our understanding of (un-) sustainability, that we do not focus our attention too heavily on conscious (or even human) thought in our discussions and analyses. The challenge involves and includes the specific settings of our perception and emotions and the agencies exercised by bodies, things, objects, organizing principles, and so on, at least as much as by conscious, language-based thought. Action is never context-free, but rather is always trapped within that tumultuous "zigzag paradigm" of ontological becoming. Perception, cognition, and emotion are learned and mastered practices that form part of the normal process of human development. The cultural specifics of those developmental processes constitute a large part of the totality of our implicitly learned relations with the world that situate us somewhere along the continuum between sustainability and unsustainability.

Once we understand the role and agency of the world to form and parent us, we gain more insight into why Cartesian mediational means are inadequate to the task of tracing Ariadne's thread out of our already-lived-in unsustainability world. The challenge of sustainability, though, becomes much more difficult when our entire psychological development takes place with and through inherited practices with an already-lived-in materiality that are in many cases psychopathological. There is, for example, a high school near my home that was designed without windows. If you ask the teachers there, most of them will agree that windows would act as a distraction for the students and divert their attention away from their lessons. Such architectures and conventional wisdoms, having once been experienced as an integral part of childhood, are difficult to dislodge precisely because of the zigzag mastering they entail. Sustainability may ultimately depend on the institutionalization and deliberative practice of organized social forgetting, unlearning, and dissociation. I see this as being to some extent unavoidable, given the historical tumult that now possesses us. But the hope is that this continuing tumult, having already deformed our available tools, might still not dictate that the project of sustainability must fail, either in its quite modest theoretical form as in the present story, or more broadly as a hoped-for societal transformation.

The world, then, is not simply something that we can learn how to read or how to extract pre-existing and implicit meaning from. The windowless school becomes meaningful within an entire cultural context. To think of the world as a container of information is simply the wrong and the unsustainable way of coming to know, and the wrong way of conceptualizing the process of coming to know the world. According to Malafouris (2013a), for example,

> Meaning does not reside in the material sign; it emerges from the various parameters of its performance and usage as these are actualized in the process of engagement. That means that the material sign engages us primarily in "pragmatic-epistemic actions" rather than hermeneutic circles .... Meaning is the temporally emergent property of material engagement, the ongoing blending between mental and the physical .... Material signs have no meaning in themselves; they merely afford the possibility of meaning, as a door affords us the possibility of being opened. In real life, to interpret a material sign is ... to become habituated with the interactive possibilities and consequences of its performance in context without bothering about exegetical questions. Material signs do not represent; they enact. They do not stand for reality; they bring forth reality. (pp. 117–118)

This passage is perhaps the most elegant way of stating that it is not true that "there is no alternative" (TINA). There are, in fact, an unimaginable number of alternatives, partly because the world is actively participating in its enactment. It is not composed of separate and lifeless inert matter waiting to be dominated by a Western-Cartesian thought. Unfortunately, Cartesian mediational means, whether windowless classrooms, financial markets and commodification, egoistic individualism, the mistaking of information for wisdom, the emphasis on conscious, calculative, quantifiable, methodologies, and so on, all militate against such a recognition. Things appear to us as they do precisely because we have been poorly parented, with a meager and impoverished historical inheritance and by a Cartesian-configured environment. And although there may be limits to the world's cooperation with the Western-Cartesian project, it does, for the most part, go along with the bad parenting induced by this *enfant terrible*. It collaborates in concrete and asphalt, televisions and iPhones, mowed lawns, recycling bins, and political boundaries. But this collaboration usually works against sustainability, whether our available conceptual tools can admit that or not.

There are many reasons, then, why the alternative narrative put forward here will not fit and find favor. One is tempted to conclude, when looking at the world, that all our mediational means have become dangerously hallucinatory. Otherwise, how could we have let the world come to this? It is not as if we did not perceive or recognize many of the problems confronting us. The more fundamental issues Malafouris and many others point to, whether semantic or substantive, are directly linked to the Cartesian perspective. From this perspective, it is difficult to resist the methodological error of searching for a single metaphysical cause with a single geographic or anatomical location—the wolf, or the structure of the human DNA molecule, for example—when addressing (un-) sustainability. It is also far easier to recognize flaws in the currently available mediational means than to invent problem-free alternatives and bring those alternatives into the world already spun as whole cloth.

Of course, many tools are available to help us to frame environmental issues from various perspectives. Sandri (2013), for example, claims that "engaging with a systems framework can transform existing ways of seeing issues and open the way for more holistic awareness" (p. 814). If only the systems perspective were adopted and mastered, this "more holistic awareness" would follow, with sustainability not far behind. But the problem is, once again, framed as a visual and cognitive incapacity that locates (un-) sustainability squarely within the human cognitive-sensory-motor system. There is an epistemic solution to sustainability that simply involves filling or implanting those brains with the requisite ("systems framework") knowledge. While Sandri is quite right that "existing ways of seeing" are problematic, we need to recognize that the materiality of the already-lived-in world is every bit as implicated in the current unsustainability as any conceptual model or cognitive tool.

The alternative Western perspective, as we have already seen in the case of Diamond's (1999) environmental determinism, looks to analyze and correct mistakes or incapacities located in the physical world. More and better domesticated plants and animals, disease resistance, biodegradable packaging, pollution controls, recycling regimes, full-cost accounting, or newer and better technologies of any kind generally involve the implicit assumption that something in environment is not performing up to standard and requires remediation of some kind. But (un-) sustainability always requires many co-conspirators, both human and nonhuman, material and ideational. The agency that gives rise to a regime of (un-) sustainability is a *distributed* agency. It is not nurture *or* nature; the two are always inextricably intertwined, and specific contexts induce, enable, and frame the exercise of agency. Rather than falling into the Cartesian trap of locating (un-) sustainability either in the brain or in the world, the perspective of mediated action raises the possibility of locating (un-) sustainability in the "mediational means" or "cultural tools" of human action (Wertsch, 1998, p. 17), and those mediational means are distributed throughout the person-tool-environment triad. We need to remember that the mediators, or cultural tools, are generally amalgams of matter and thought, much as Vygotsky's (1978, p. 51) example of the knotted handkerchief that reminds its owner to buy bread on the way home from work is to be understood as a person-handkerchief-reminder practice. The handkerchief as a mediational means for memory is no longer simply a material entity; it is material and mind extended and transformed according to their co-constitutive role in a certain practical setting.

Materially Engaged (Un-) Sustainability

Material Engagement Theory provides us with a way of thinking about the kinds of mediational means we have been discussing. This approach is a recent development, and the term has not been universally adopted even within its originating academic fields, most notably social and cultural anthropology and Paleolithic/Neolithic archaeology. In 2001, the archaeologist Colin Renfrew published a book chapter outlining a framework for an "engagement theory." The chapter "Symbol before Concept: Material Engagement and Early Development of Society" (Renfrew 2001, pp. 122–140) did not set out to offer a completely new theoretical framework but does outline a little of the history of the emergence of a focused discussion on material engagement. He mentions in particular the influence "of attending a symposium whose discussions were published under the title *The Social Life of Things*" (p. 123). This book (Appadurai, 1986) in turn has been widely cited and influential in other areas, such as social theory. Yet these other works do not provide the fully formed theoretical framework for understanding the inextricable entangling of psychological development with materiality offered by MET. Renfrew's thesis is essentially that material engagement gave rise to the modern human species and not the other way around; that is, it was the affordances of material tools and artifacts that created the evolutionary context for the emergence of our contemporary species.

The earliest mentions of the term *Material Engagement Theory* that I am aware of appear in separate chapters written and published by Renfrew (2001, pp. 23–32) and Malafouris (2004, pp. 53–62). The latter chapter can be found in *Rethinking Materiality: The Engagement of Mind with the Material World*, edited by Desmarrais, Gosden, and Renfrew; this book very productively brings together several scholars who explore the ways in which "the concepts of engagement and materialization bring attention to material culture and to the activities and behavior by means of which humans interact with and construct their social worlds" (p. 1). Most, if not all, of the authors in this edited collection address the same issues that MET seeks to address. Boivin's chapter (pp. 63–72) in that publication, for example, represents an attempt to "collapse the mind-matter dichotomy" (p. 63). In 2013a, Malafouris would publish the first comprehensive manuscript on MET as *How Things Shape the Mind: A Theory of Material Engagement*. In the essay, *Mind into Matter*, also published in 2013b, Malafouris claims to have "developed—material engagement theory (MET)" (p. 28). In this short essay, Malafouris concludes,

> MET offers a new way of understanding how different forms of material culture, from the stone hand-axe to the iPhone, may have provided a powerful mechanism of defining, but also transforming, what we are and how we think. Mind-changing technology has a futuristic, sci-fi ring to it, but what most people don't realise is that humans have used it since they first evolved. (p. 29)

Although MET is now mainly associated with Malafouris, that association certainly reflects the mediational means of the "copyright age" (Wertsch, 1998, pp. 17–21) more than any creative capacity on the part of Malafouris to invent MET *ex nihilo*. In fact, MET's origins can be traced across a variety of disciplines, research programs, and scholars. Malafouris (2013a, pp. xiiv-xv) acknowledges immediate colleagues who played an important role in the development of the theory, including Renfrew (2003), Knappett (2005), Gosden (1994), Boivin (2008), Clark (1997), Donald (1991), Hodder (2012), Hutchins (1995), Wheeler (2005), and Ingold (2000). The opportunities provided by university infrastructures, including academic conventions; archaeological discoveries; research protocols and methodologies; and the dissemination of ideas through conferences, personal conversations, and publications also undoubtedly contributed to the development of MET. As already indicated, the wider ecologies of ideas giving rise to MET might be traced through many other theorists such as Burke (1966), Bateson (1972), Bourdieu (1977), Vygotsky (1978, 2012), Wertsch (1991, 1998), Tilley (1991, 1994, 1996, 1999), and Bennett (2004, 2010) among many others. The extended mind theory of philosophers like Clark (2003) has played a particularly important role in providing material criticisms of Cartesian notions of a mind detached from an "external" world (Malafouris, 2013a).

There are, in other words, several scholars working in the field of cognitive archaeology whose work could be applied directly to sustainability issues and research questions, regardless of whether they use the term *MET* or not. For example, Boivin's 2008 book, *Material Cultures, Material Minds: The Impact of Things on Human Thought, Society, and Evolution*, makes no mention of either

Malafouris or MET, despite the acquaintance of the authors and the general agreement and overlap of subject matter. Boivin's work also brings us to a post-Cartesian perspective and provides enormous potential for sustainability studies. For example, Boivin states, "I will attempt to show that the history of human engagement with the material world is not so clearly one of mind being imposed on matter, or form on substance, but rather a history in which mind and matter, and form and substance, continually bring each other into being" (p. 23).

A more recent paper, *Creativity, Cognition and Material Culture* (2014), co-authored by Malafouris, Gosden, and Overman, describes another facet of Malafouris's contribution:

> Drawing on his theory of material engagement, in which cognition is envisaged as the dynamical interaction of brains, bodies and materiality, Lambros Malafouris proposes and examines the notion of creative "thinging," as the human species-unique commitment to the discovery of new varieties of material forms through a saturated, situated engagement of thinking and feeling with things and form-generating materials. (p. 4)

MET thus also overlaps with the work of phenomenologists working with "thing theory" (see, for example, Candlin & Guins, 2009). The main point here is that the subtle differences between these research methodologies are miniscule in relation to the powerful advantages they might provide in bringing our understanding of sustainability out of the Cartesian delusion.

In any event, the term *MET* provides a useful shorthand for a collection of ideas that are extremely relevant to our understanding of the challenge of sustainability. Perhaps sustainability theorists will ultimately choose a different nomenclature—"postdelusion studies" springs to mind—but the comprehensive theory outlined by Malafouris (2013a) has the advantage of providing a manifesto that enables sustainability theorists to hit the ground running.

It should not be surprising that MET has emerged from work in the field of cognitive archaeology, where connections between material artifacts and mental processes, especially in terms of toolmaking, have been a primary focus. As Malafouris notes,

> The beginning of early-hominid stone flaking ... may have been an important threshold in the prehistory of mind .... No other species has been or can be defined *as a species* on the basis of its relationship with tools and material culture. (p. 153; italics in original)

Malafouris provides an extended account of stone knapping arguing that, far from representing the externalization of a mental plan to produce a material object, knapping itself is the mediational and material process by which the human mind emerges. Knapping is an embodied and materially engaged process whereby "the knapper first thinks through and with the stone before being able to think about the stone and hence about himself as a conscious and reflectively aware agent" (p. 176).

The preliminary question MET seeks to address is, "What's the matter of which our minds are made?" (Theiner & Drain, 2017, p. 837) As Malafouris (2004) explains,

> There is no mind apart from the world (though there might be a brain), just as there is no University apart from the colleges and departments. The analogy may be crude but it makes the basic point: mind is immanent on the one hand in the sequence of technical gestures required for the production of the tool and the various media that are brought to bear on that sequence, and on the other in the skillful actuality that transforms the tool into an agent imbued with cognitive and social life. (p. 60)

Within the MET framework, in other words, there is no ontological separation of mind and matter, and no hierarchical ordering of mind over matter. The mind "extends" into environment at any given moment of cogitation just as the world "extends" into the human mind recursively (Abel, 2015; Hundert, 1997).

These insights are, I would claim, directly applicable to sustainability theory. Malafouris (2013a), building on the parity principle, makes the point that "in a very important sense, then, from the perspective of material engagement, *cognition has no location*. The active mind cannot be contained. Cognition is not a 'within' property; it is a 'between' property" (p. 85). Sustainability is likewise, I would say, without location. This between affects both cognition and the world's material forms. That in-between relation also materializes into a trace, just as the child playing with the bobo doll retains a trace of his or her interactions in mind even after the child-doll assemblage ends. In this way, it becomes an obfuscation to attempt to disentangle and conceptualize mind and matter as distinctly separate entities. Like MET, the understanding of sustainability set out here "wants to change our understanding of what minds are and what they are made of by changing what we know about what things are and what they do for the mind" (p. 141). While what a mind is might be partly an ephemeral in-between, it is also partly a force within a material ecology that "cannot be contained."

At the micro level too, MET can be employed to shed light on the *intention* behind (un-) sustainability. Following Malafouris (2008, pp. 19–36), MET works to problematize the conceptualization of (un-) sustainability as being pre-designed in the brain prior to being carried out with bodies and tools in environment. By paying attention to the origins of intention, we arrive at a clearer understanding of the *agency* of (un-) sustainability. As Malafouris (2008) explains,

> Accepting that agency is about causal events in the physical world rather than about representational events in our mental world, it follows that if an association between agency and intentionality can be made, it has to be with the type of intentionality here called "intention-in-action." (p. 30)

Although Malafouris's model of intentionality specifically builds on, and critiques, the analytic philosophy of Searle (1983), the same conclusion can be arrived at by combining the work of Wertsch's social-psychological discussion of "action" (as outlined in Chapter 1, pp. 7–12) or Heidegger's phenomenological concept of "being-in-the-world." The point is that the world plays a role in disseminating human intention, and this means that agency tends to be distributed. When our perspective shifts from an atomistic conceptualization of intentions originating *ex nihilo* in the brain to a more web-like perspective of something approaching an infinitely connective cosmology *and* that new perspective is evidence based and might be explained in an easily comprehensible narrative form, then the possibilities for sustainability increase exponentially.

MET represents a truly refreshing interdisciplinary methodology focused on answering important questions rather than policing disciplinary boundaries. Of course, MET is not a panacea, and any and all theories, including MET, are still likely to be misappropriated in ways that revert back to the dominant Cartesian (and capitalist) norms. It should perhaps come as no surprise that the early applications of MET include marketing, branding, and for-profit design (Bean, Figueiredo, & Larsen, 2017). But MET will have implications across all fields of knowledge and offers an opening on solutions to sustainability problems that could be a critical part of the long and slow escape from Cartesian-based delusion, regardless of these initial misapplications. Once MET is applied to the task of sustainability, the possibility of alternative minds and other worlds is opened.

The possibilities of MET succeeding in this respect depend on how its prognosis might become easily transmissible within a comprehensive narrative as "a recursive vision" (Harries-Jones, 1995). And MET *is* able to explain the processes of sociocultural-ecological change through time as changes in the qualities of such cognitive, embodied, and materialized relationships from the Paleolithic emergence of the human mind to the current state of global, but by no means universal, dystopian unsustainability. As we have seen, Malafouris (2013a) has argued convincingly that stone tools acted as "cognitive prostheses," (pp. 154–5) providing new dynamic opportunities for socio-psychological development, and he has already applied MET to wide variety of other tool-using scenarios, such as modeling clay, using it to move our perspective beyond "a sterile neurocentrism" (p. 169). If, where the common hand axe of the Neolithic period is concerned, "the material qualities of the stone make a causal contribution to the spontaneous implementation and realization of the plan that brings forth the tool as an artefact" (p. 177), then it is not unreasonable to suppose that today the plastics, tarmacs, concretes, antibacterial agents, pharmaceutical alchemy, and other "material qualities" of environments continue to co-constitute our collective cognitive and emotional processes, bringing forth the entire world "as an [unsustainable] artefact."

MET helps us explain why (un-) sustainability is not something that can be located, examined, and manipulated; rather it is distributed across the forms of relationship between brain, body, places, tools, and environment, and always in co-constitutive and reflexive, zigzag interplay. It elucidates why sustainability is not simply a problem of knowledge or information but is thought, felt, and practiced by the entire planet, including atmosphere and ocean, wolverine and orchid, and pavement and chalk. It resonates through the guy traumatizing the neighborhood with a leaf blower at 7 a.m. on a Sunday and the business executive who is more concerned about her next automobile purchase than the ecological consequences of her routine business decisions. It tars every leaf and twig, every fish and fowl, every thought, every wish, every invasive species, or genetically modified laboratory rat along the way. The employment of MET offers a very different analysis of sustainability than those historical narratives that have dominated sustainability research and policy development since the beginning of the environmental movement.

We might, for example, think of the story of the "Tragedy of the Commons." This model of (un-) sustainability has been particularly resilient within the social imaginary, but like mainstream economic theory, it becomes real only because "we become what we pretend" (Bateson & Bateson, 1987, pp. 167–182). In this case, the pretense is one of a Newtonian model of self-interested subjects and mechanistic objects, whereby, as Tsoukas (2005) explains,

> ontologically, a phenomenon is supposed to consist of discrete, objective elements, whose law-like associations the analyst will identify through the construction of an abstract model, for the purpose of predicting and, if possible, controlling the phenomenon at hand. The Newtonian view assumes an objectivist ontology, works with mechanistic epistemology, and enacts an instrumental praxeology. (p. 213)

The *tragedy of the commons*, a term initially coined by Forster Lloyd in 1833 to describe the resource collapse of over-grazed common pasturelands, was famously popularized by Hardin in his 1968 essay of the same name. This story is entirely dependent upon a shared understanding of people as "discrete, objective elements" whose behavior is inherently and unchangeably selfish and strictly self-referential. The tragedy of the commons is also a classic "abstract model" lacking any capacity for reflexivity. It is meant to be a description of an undeniable and unalterable reality, but has itself helped form the "ideas" or cultural/cognitive tools used to interpret any event or substance across time and space (see Galbraith, 1958/1998, pp. 6–17). By contrast, the MET model allows us to perceive ourselves as person and tragedy-of-the-commons–conceptual tool, rather than as inherently self-interested, profit-maximizing neurotics. From the MET perspective, people can be seen to be mastered by such established practices as overexploiting common resources *and* understanding themselves as rationalistic self-interested individuals. The tragedy of the commons is, in other words, far from being an immutable law of nature, but rather results from mediational means coupled with ecological ignorance, stupidity, or possibly some form of Western knowledge-induced insanity embedded in them.

Other dominant perspectives, like the systems-theory approach promoted by Sandri (2013, see Chapter 2, p. 54) take us partway out of Newtonian madness, but do not entirely extricate us from its ill effects. Boulding's work (1978), for example, contributed to the development of systems theory and his account of "ecodynamics"; places things and ideas on a nearly equal footing with each other; and also exhibits an explicit awareness of a fluidly changing entirety of brains, bodies, tools, and places (see also Berkes, Colding, & Folke, 2003; Berkes & Folke, 1998; Berkes & Ross, 2016).

Bateson (1972/2000), a pioneer in systems thinking, or what he termed "cybernetics" (and of course of the ecology of mind) was cautiously optimistic about the promise of the systems perspective:

> I included cybernetics as the second historic event in my lifetime because I have at least a dim hope that we can bring ourselves to use this new understanding with some

> honesty. If we understand a little bit of what we are doing, maybe it will help us to find our way out of the maze of hallucinations that we have created around ourselves .... I think that cybernetics is the biggest bite out of the Tree of Knowledge that mankind has taken in the last 2000 years. But most of such bites out of the apple have proved rather indigestible—usually for cybernetic reasons. (pp. 483–484)

Bateson's systems perspective differs markedly from Hardin's tragedy of the commons, and Bateson also recognizes that knowledge types exercise forms of agency. As Popkewitz (2017) reiterates,

> Cybernetics theories connect to systems thought. Systems are an abstraction to actualize future society and people; the abstraction embodies principles that are not empirically deduced but are *a priori* and *self-referential and self-authorizing*; that is, its mode of ordering and classifying inscribes internal boundaries in defining problems, contexts and the possibilities of change. This is not unique to cybernetics as a system of reason. (p. 5; italics in original)

Advocates of systems approaches to sustainability tend to emphasize its novelty and the recursive nature and adaptability of the knowledge it produces:

> Facing complex co-evolving systems for sustainability requires the ability to cope with, adapt to, and shape change without losing options for future adaptability. The irony is that the mental model of optimal management of systems assumed to be stable and predictable has in many respects reduced options and removed the capacity of life-support systems to buffer change. The insurance for dealing with the unexpected has been driven down by suppressing disturbance and reducing the diversity of the environment .... *A fundamental challenge in this context is to build knowledge and incentives into institutions and organizations for managing the capacity of local, regional, and global ecosystems to sustain societal development in the context of uncertainty, surprise, and vulnerability*. (Berkes, Colding & Folke, 2003, pp. 353–354: emphasis added)

However, the authors, while rejecting the Newtonian assumption of a manageable, "stable and predictable" world, still assume the objectivist, epistemological, and ontological stance of the doctor applying complex knowledge to exterminate the contagion of unsustainability. Rather than suppress disturbance, the authors would allow natural cycles of variability to increase resilience. (This might, for good reason, sound much like the faith expressed in market mechanisms through the magic of "creative destruction" (Schumpeter, 1934/2005) to always optimize the world into a Panglosian utopia (Voltaire, 1759/2007)). From the systems perspective, the patient does at least include both physical (ecological systems) and human (social systems) landscapes, but otherwise remains fundamentally unchanged. The conceptually disconnected/context-free human brain largely retains its imagined sovereignty.

In focusing almost exclusively on objects, Western knowledge's cognitive tools, including systems theory, demonstrate a scotoma or blind spot, hiding or obscuring the characteristics of relationships. Knowledge, for example, is not like a software program that is operated by the hardware of a brain. Brains leak. Brains and bodies become accustomed to specific practices. One way systems thinking *could* represent a detour from the Western-Cartesian heritage would be to assign agency in a distributed way, not just conceptually but also in practice. This would change the nature of the relationship between, say, people, forests, and fire, from one where people impose fire-practices upon the forest to one where a partnership is negotiated between people, fire, and forests. This would involve an actual, and not simply a conceptual, redistribution of agency.

Sustainability is not simply about finding the most accurate representation of reality within the perceptual, emotional, and cognitive parameters of the human organism. We must also ask ourselves how well does our knowledge system perform, over the longer term and within a broader ecology of mind/matter, as a medium for the achievement of human well-being? The assumption that systems thinking is a magic bullet to solve the tragedy of the commons seems, at best, naive. We are well past the point where a little tinkering, complemented with a few new parts, could transform our situation into something that could even remotely be considered sustainable.

Pessimism aside, MET strongly suggests that, as developmental psychologist Bruner (1986) argues, "many worlds are possible" (p. 149). Bruner's work substantially influenced the development of MET. Building on Vygotsky, Bruner focused on mediated action to improve our understanding of processes of psychological development and learning. His utilization of "surprise" (p. 46) as a methodological tool is characteristic of the insights MET can offer to any discussion of mind and to our understanding of (un-) sustainability: "Surprise is an extraordinarily useful phenomenon to students of mind, for it allows us to probe what people take for granted. It … is a response to violated presupposition." Although Bruner did not conceptualize mind as environmentally extended and distributed, his insights still apply. Surprise, though, is not something that just occurs in a brain, but rather is distributed and relational. This point is critical for both MET and sustainability theory.

System-theoretical approaches to "environmental management" also recognize surprise, as Holling, Berkes, and Folke (1998) make clear:

> Surprise is inevitable. We use the term to denote the condition when perceived reality departs *qualitatively* from expectation. Surprises occur when causes turn out to be sharply different from what was conceived, when behaviours are profoundly unexpected, and when action produces a result opposite to that intended. (pp. 346–347; italics in original)

However, the systems perspective frames surprise as the consequence of the inevitable inadequacy of human knowledge in relation to a dynamic and complex world. Bruner, by contrast, saw world making as a process linked directly to specific sets of cultural tools. "Perceived reality" is always, from Bruner's perspective, a *learned* reality in the sense that perception is a mediated action. The metaphor that observes that environment is like a parent would have, according to Bruner (1986), the capacity to open another "possible world." Mediational means, in other words, always entail peoples and environments as entire worlds in action.

The disaggregated perspective of Cartesian atomism separating mind from matter fails to deliver sustainability. Although it delivers much surprise, (even "*predictable* surprise"; Bazerman, 2006, p. 179), clearly indicating that it is an extremely poor representation of reality, this perspective also exerts an addictive power in the form of a system of presuppositions that (so far) seems unbreakable. Cartesian atomism exerts unparalleled influence on the territory it maps. That influence leads us ultimately toward the demise of both the territory and the cartographer, while the maps' presuppositions remain beyond question. Catastrophic trends become obvious to anyone with senses, and yet the obvious ongoing collapses fail to interrupt established routines and institutionalized practices. To act in the world, according to the Cartesian perspective, always involves a plan hatched in the brain and *then* imposed on a passive and receptive tabula rasa, regardless of the mountains of evidence that suggest this template sooner or later always fails.

Cartesian atomism enables, supports, and maintains the unsustainability mind (see Chapter 1). Every time I think of myself as separate and distinct from the world, that thought materializes in the synaptic structures of my being. Every time I conceptualize myself as a utility-maximizing calculation machine whose primary function is to accumulate utility in the form of financial wealth, I impoverish myself, my world, and foreclose other potential futures. The map or mental model, in other words, bends to conform to the subconscious workings of socioculturally directed patterns of material engagement. We may find ourselves, almost like automatons, enacting implicitly learned patterns of tacit knowledge. Deforestation (Harrison, 1992), French formal gardens (Weiss,

1995), and suburban lawns (Robbins, 2007) provide three examples of brains, bodies, and tools together driving whole environments insane through the nature of relationships they form (Bateson, 1972/2000, pp. 486–492). Those forms then continue to spread insanity in widening and deepening loops of dysfunction around brains and bodies, to places and back again.

Perhaps it seems easier and more practical to transform physical places and material artifacts than to change the cultural tool set, the perceptual, emotional, and cognitive repertoires of the "noosphere." Boulding (1978) defined this as "the totality of the cognitive content, including values, of all human nervous systems, plus the prosthetic devices by which this system is extended and integrated in the form of libraries, computers, telephones, post offices, and so on" (p. 122). However, in truth, systems of materialization and systems of meaning must transform simultaneously and the "cognitive content" of Boulding's noosphere concept should be extended in conformity with MET to include such things as funerary monuments, highways, oceans, skyscrapers, trees, pesticides, and virtual-reality machines—in short, anything that becomes integrated into the workings of a human mind. Our inherited tool set nearly always prescribes the concept of having brains take better control of environments, without recognizing that "control" is something that goes at least as much in the other direction. The world always bites back, so what matters is the form of relationship established and put into practice. A relationship modeled on a dominant-submissive mediator will bring into becoming a very different world than a relationship modeled on parent-child co-development.

One might describe sustainability then as something that could result from the orchestration of a sort of harmony between all the component parts of embodied, distributed, mediated, situated, extended, and enacted action, including brains, bodies, material objects, narratives, meaning systems, biogeochemical processes, the laws of physics and chemistry, and so on. While true, this description is still far from complete since there is a sense in which there is always a sort of correspondence, a bringing into agreement or isomorphy of the parts into a whole that emerges from any human activity. Even the neoliberalized social mind of unsustainability is brought into a kind of alignment with environment in the sense of having coordinated brains, bodies, and environments within a self-reproducing system, even if ultimately the system represents a dead end. Unfortunately, the apparent direction of movement of that system would seem to be the extermination of beauty, if not the ultimate extermination of the human species. This is clearly not the kind of harmony that sustainability requires.

Toward a Materially Engaged Partnership Mind

So far, I have argued that MET draws upon critical aspects of previous approaches to provide new insights and understandings of our (un-) sustainable mediational means, whether expressed in the form of the tragedy of the commons or of systems thinking. In each case, putting mind and matter back together and then structuring that reunification with the parent-child metaphor brings surprising results. We might, (as we shall see later in Chapter 4) also apply this directly to the dominant economic paradigm of neoliberalism, which, according to Ferguson (2009), can be described as follows:

> The key elements … always include a valorization of private enterprise and suspicion of the state, along with what is sometimes called "free-market fetishism" and the advocacy of tariff elimination, currency deregulation, and the deployment of "enterprise models" that would allow the state itself to be "run like a business." (p. 170)

Neoliberalism encapsulates and exacerbates many of the key features of Cartesian mediational means, such as isolated, self-interested, and rationalistic individualism. Certainly, it also involves more than the Galbraithian (1958/1998, pp. 6–17) formulation of a set of ideas used to interpret events. It has been produced within and continues to produce a materially unsustainable world where all alternative metrics are delegitimized. For example, any emphasis on aesthetic harmony (Berleant, 1994), or beauty, or what Bateson (1972/2000) described as "grace" fails to resonate within a paradigm of profit maximization that only focuses on human action or separate physical entities.

We can all agree that within a given reality certain explanatory theories will resonate more than others. The question of *why* one theory resonates is more contentious. Where minds are considered dominant and matter submissive, people are sorted and hierarchically arranged according to their capacity to dominate matter. Neoliberal explanatory models will, *ceteris paribus*, be more likely to resonate in such circumstances. But if sustainability resides in the patterning of how minds weave and are woven from worlds, then this clearly presents a model that is unlikely to resonate with the epistemological assumptions of Western knowledge systems generally and neoliberalism specifically.

Since neoliberalism operates ontologically as much as epistemologically as an entire person-tool-environment matrix, it is not a single billiard ball that can simply be taken off the table. Its influence is pervasive. For example, where basic-income policies have been implemented, their success is evidenced in a reduction of material want, inequality, and a broad range of social problems. This, however, does little to make the policy politically sustainable over the longer term since it does not resonate with the neoliberal agenda, and this is not remediable with more information or by conferring critical thinking skills. The counter-narrative of basic income is forced to contend with a rhetorical and an entirely neoliberalized material environment (see, for example, Robbins, 2007). We will return to these problems associated with economism and neoliberalism in more detail in Chapter 4.

Similar problems exist in educational literature addressing sustainability. If sustainability is theorized as a problem of mind, such as a problematic narrative, model, or idea, then it would seem as though education would be the correct locus of remediation. But mainstream education is rarely focused on developing an environmental aesthetic or bringing about harmony, or "grace" within a process of material engagement. Education tends to operate within a dualistic ontology where knowledge is understood as being unproblematically quantifiable. MET, applied to education for sustainable development, would at least bring materiality into the educational design process and problematize the conceptualization of knowledge as cumulative.

Mezirow's (2000) transformative learning theory (TLT) has been viewed by many as a promising way forward in the quest for ecological literacy and sustainability (see also Calleja, 2014, for a good account of transformative learning in practice). The transformation of TLT has been well explained as follows.

> Mechanisms for facilitating TL include the following key phrases connected to the learning process: 1) a disorienting dilemma; 2) critical questioning and self-reflection; 3) sharing of one's transformation process; 4) exploration of new roles and actions; 5) planning a course of action; 6) provisional trying of new roles; 7) building competence and self-confidence in new roles and relationships; and 8) a reintegration into one's life directed by one's new perspective. (Bushell & Goto, 2011, p. 1245)

As we have already seen, disorienting dilemmas (or surprises) are not in short supply. But, as the subsequent phases of Bushell and Goto's model suggest,

TLT is primarily directed at the internalized operating code of the student, and societal transformation toward sustainability is conceptualized according to underlying and implicit assumptions about individual human agency located in brains operating with varying amounts of propositional (i.e., Western scientific) knowledge. The process of TLT is thus often conceptualized as the deliberate removal of some bits of knowledge from brains (steps 1–3 above) and the replacement of the deleted knowledge with other bits of knowledge (steps 4–8).

If we apply MET then the process of transformation becomes much more complex, but this might also explain why TLT applied to education for sustainable development has had such meager and generally compartmentalized successes (Lozano et al., 2015). MET would suggest that transformative learning and sustainability cannot simply emerge from ideas, but must also be reflected in places and tools. Disconnected knowledge is unlikely to result in a changed way of becoming in the world that would represent real sustainability. For example, forest-management skills, while having value in the labor market, will continue to frame forests as wood products and commodities. Urban-planning skills will continue to operate on two-dimensional maps of geographic communities conceptualized essentially as tabula rasa. Until the land is assigned some degree of agency, both as a conceptual model *and* as a biosemiotic materialization, progress toward sustainability is likely to be muted, at best. Societal transformation toward sustainability is about creating atmospheres of sustainability that will entail new roles and relationships, new competencies, and new practices entangling new materializations, as well as new perspectives and identities. The missing ingredient is some form of guidance out of our zones of proximal development (ZPD), operating as the "double binds" (Bateson, 1972/2000, pp. 206–207) of our collective historical inheritances. Vygotsky (1978) defined the ZPD as "the distance between the actual developmental level as determined by independent problem solving and the level of potential development as determined through problem solving under adult guidance or collaboration with more capable peers" (p. 86). That guidance does not need to be structured in the form of "ideas" since these remain largely impotent until they come to be materialized in artifact and institutionalized in material cultural practices.

How then are we to create a platform upon which to build sustainability and an ethics of sustainability? As we have seen, Material Engagement Theory incorporates themes found in many other fields, including the cybernetics of Bateson, cognitive science, social psychology, anthropology, and phenomenology; taken together, these might form a workable foundation for a research methodology for sustainability. Phenomenology, for example, has already been explicitly applied to questions of sustainability. This passage from Stefanovic (2000) illustrates the overlaps between MET and phenomenology:

> If human being-in-the world is, as Heidegger describes it, a fundamental temporalizing and spatializing activity of emplacement, it is clear that places within which we find ourselves are hardly incidental backdrops of our everyday existence. Instead, they virtually define who we are by conditioning our moods, our sense of meaning, and orientation in the world. Rather than merely providing physical shelter or avenues for community development, a holistic, phenomenological description of places reveals that they are the primordial domain of our embodied existence. (p. 112)

This phenomenological perspective underscores the constitutive role of materiality and implicitly recognizes the role of mediational means in those constitutive processes.

There are, however, a few small caveats that need to be laid out with respect to phenomenology. First, without an explicit focus on mediational means, phenomenology runs the risk of smuggling some aspects of the Western-Cartesian mediational means back into its products and prescriptions for sustainability. While environment is unquestionably the "primordial domain of our embodied existence," it is also important to point out environment's role as a bank of mediational means and as a constitutive aspect of mind. Second, phenomenology is not always as attentive as it might be to the "bulk of life" in terms of the socio-historical development of particular mediational means. This might mean, for example, that we might potentially be led astray by failing to scrutinize a mediating conceptual tool such as "efficiency." Stefanovic (2000), quoting from the "Afterward" in editors Ann Buttimer and David Seamon's *The Human Experience of Space and Place* (1980), notes that this possibility is apparent, for example, when "David Seamon suggests that 'rootedness in place promotes a more efficient use of energy, space and environment than today's place relationships, which emphasize social mobility and the frequent destruction of unique places'" (p. 115). Phenomenology is perhaps less focused on *how* specific mediators, like efficiency, come to master us or on asking what might replace those tools once they are recognized as deficient. The "efficient use of energy, space and environment," for example, aligns with a shared understanding of sustainability as human epistemic mastery and environmental management.

Where MET can add something meaningful to phenomenology is in its focus on ways in which relationships get structured through participation in communities of practice and how mediational means form and reform in action. But phenomenology does have at least one important advantage on its side: It recognizes that we are thrown into a pre-existing world that accommodates us and to which we become accommodated in various ways. MET emphasizes the materiality of this accommodation. There is no neutral way of being in the world, no way to be born and develop into a human being, or any other living thing for that matter, without being squeezed through the interstices, acquiring and being acquired by the specific mediational means available, in the already-lived-in world. For most of life on earth, the process of learning is nearly synonymous with that of simply being alive. Perhaps the plants we encounter can most visibly be understood to always accommodate themselves to their surroundings, and we should not doubt that even the roots of a given plant in a flower pot will develop differently from the same roots in a field of flowers or in a forest.

The concept of tacit knowledge confronts us with the point that much of our cultural tool set escapes our conscious awareness. Our mediational means become, through normal human developmental processes, our second nature, the nature we master and that masters us. For the most part, these relations are sunk below conscious awareness as indicated in Bateson's (1972/2000) essay entitled, "Style, Grace, and Information in Primitive Art." As Bateson points out,

> The lions in Trafalgar Square could have been eagles or bulldogs and still have carried the same (or similar) messages about empire and about the cultural premises of nineteenth century England. And yet, how different might their message have been had they been made of wood! (p. 130)

The ability to read the meaning of stone lions is inherited as part of a socially shared tacit knowledge, acquired and learned implicitly. Replacing a Cartesian with a relational ontology tacit knowledge is a bit like trying to replace the stone lions in Trafalgar Square with wooden ones. It first requires that we recognize what we had previously been largely unaware of regarding the mediational role of the stone and not just the lion's surface imagery and symbolism. Like the meaning of a mowed lawn (see Chapter 6), everyone can understand it at a subconscious level but has difficulty explaining it in words, but then words are seldom required to "know" what the stone lions or the mowed lawn means (see also Lakoff & Johnson, 1980).

Materiality imbues our world with meaning. Some meanings can be traced to the physicality of the body and its positionality in the world (we tend to think, for example, of time as moving forward, or love as being warm, meanness as cold, etc.), but others arise more indirectly:

> Rain water marks the rhythm of the passing seasons and the blossoming of berries and fruits, with every society in north-western Europe experiencing meteorological fluctuations. With respect to some of the attributes of water, be it from the sky or the river, we should note that associations with life and liquid can transcend time, space and society—the politics of liquid .... Contemporary Irish examples of beliefs operating in a similar manner include the wetting of the Blarney Stone, Cork, with saliva via the mouth as a process of bodily activation (i.e., the generation of eloquence or having the gift of the "gab"). In another example, the Mandari of Southern Sudan make associations between life, fertility, women and water, and these are demonstrated by the euphemism for pregnancy, which is termed *sipi I pele* "water in the belly," while the word for semen is *kula yunusi* "making water which promotes conception." A modern structural approach to this might suggest that "wet" and "dry" as an oppositional pair mean "life" and "death." Therefore liquids are living, whereas drying is dying. (Cochrane, 2008, p. 169)

Water, lions in Trafalgar Square, or the lawns of suburbia do not exist in vacuums. We have all been acquired by one or another entire system of meaning; much more comprehensive than language, that system enables us to make implicit sense of nearly any event or thing. Knowledge of visual meaning begins to be acquired at birth, with a focus on faces of caregivers and on learning to read facial expressions. The first language we learn is generally body language. Normal psychological development is not simply about the cultural/cognitive tools we use consciously; rather, and more importantly, this development includes the tacit cultural schema that sets up our subconscious model of subjectivity, the cumulative sensibilities that master us into individual actors. The mundane events and things of routine existence work reflexively to establish and maintain a bank of tacit knowledge, while any newly encountered event or surprise becomes interpretable through extrapolation, always by reference to the same pre-existing meaning system. That meaning system, in turn, operates in a zigzag fashion, partly as the physicality of the body—the materiality of things and objects—and partly as mastered and mastering cultural/cognitive tools and schema.

Bateson (1972/2000) elaborates on this point: "Since the level to which things are sunk is characterized by iconic algorithms and metaphor, it becomes difficult for the organism to examine the matrix out of which his conscious conclusions spring" (p. 142). Bateson's solution to this problem of a seemingly inaccessible system of meaning is to work backwards:

> The question has been: Does the art tell us about what sort of person made it? But if art, as suggested above, has a positive function in maintaining what I called "wisdom," i.e., in correcting a purposive view of life and making the view more systemic, then the question to be asked of the given work of art becomes: What sorts of correction in the direction of wisdom, would be achieved by creating or viewing this work of art? (p. 147)

Here Bateson is getting at the way what we make in turn makes us and the way that what we design, once in the world, continues to design (see also Fry, 2012).

Bateson claims that art is indexical of an underlying sociocultural system of meaning that can provide information about its makers and, as Gell (1998) suggests, also about the persons who encounter the work of art. Bateson claims that "art" includes both the process of making art and the material artifacts and the effects on others resulting from that process. The art's participant, the art piece, and the originating context or culture are indexical of each other. But art is certainly not the only human activity that reflects its originating context. Any designed and/or designing process, from the building of skyscrapers to the production of econometric models is indexical of, to some extent, the entire propagating context including, among other factors, the particularities of bodies, tools, narratives, processes, agencies, memories, and atmospheres co-constituting that context.

Bateson's claims about art, then, can be greatly expanded to include anything that contains a design component. If, for example, I turn on the television and the evening news states that health-care costs are rapidly rising, I have any number of cultural/cognitive tools at my disposal to weave that information into my conscious and subconscious way of becoming in the world and for it to alter my ways of acting in the world. I constantly need to design and redesign the auto-narrative I use to organize and give structure and meaning to my life. In any given society, there will be a distribution or field of possible techniques for accommodating such news. The pattern of that distribution, and the distribution of all other design processes undertaken by members of the given society, make that society recognizable as a more or less coherent and predictable whole.

This broadened model of art and design includes both conscious and subconscious selective processes working to form material artifacts, patterns of behavior, and interpreting schemata or mastered and mastering systems of meaning. Much of these design processes operate subconsciously, simply by inducing or exercising degrees of focused attention. The mechanics of this process are nicely summarized by Cleeremans and Jimenénez (2002).

> Mounting evidence suggests not only that the brain is far more plastic than previously thought, but also that the effects of learning can be tracked all the way down to the organization of local connectivity. To wit: Expert string players exhibit larger-than-normal areas in the somatosensory cortex dedicated to representing input from the fingering digits. Likewise, not only is posterior hippocampus—a region of the brain involved in episodic and spatial memory—enlarged in experienced taxi drivers compared to subjects who do not have extensive experience in memorizing complex maps, but the observed size differences further depend on the amount of driving experience. There is also considerable evidence that the brain can recover in various flexible ways after trauma, and even suggestions that the very organization of the somatosensory cortex … depends on prenatal sensory experience. More recently, suggestive evidence for neurogenesis was also found in humans, a finding that overturned decades of unquestioned—but, as it turns out, erroneous—assumptions about the lack of regenerative cellular processes in the adult brain. These often spectacular findings all reassert that adaptation plays a fundamental role in cognition, and that its effects can be traced all the way down to the manner in which specific neural circuits are organised. (p. 3)

Our ways of acting in the world are etched into brain, body, and environment. Brain plasticity, in other words, involves the particulars of places acting to design the particulars of persons-in-places and orchestrating the weaving of persons and places into co-mingling, co-constituting, and co-determining units of analysis.

While the quantum physicist can operate with such ambiguous and slippery units of analysis that undermine conventional wisdom, social scientists are not so fortunate. The reason for this discrepancy is simple. Quantum physics can be ignored because it is not immediately apparent to the general public what implications it has for the running of their day-to-day lives. For example, the fact that the physical space of my body shares matter in the form of subatomic particles with the desk I am working on has little or no impact on the models I use to understand economy, ecology, or even the assumed boundary between me and not-me. On the other hand, to suggest that the answer to the question, "What is a human?" floats somewhere between the nonhuman bacteria in the intestine and the architectural style of one's shelter might set off a carnage among the Cartesian sacred cows that leaves people feeling hopelessly lost. If I cannot be certain what it means to be human, then how will

I know which brand of low taxes, strong military, more-economic-growth politician I should vote for in the next election? How will I know how I should feel about plastic, poverty, pollution, pornography, or pay-per-view television if my identity starts spilling out and dribbling all around, comingling with environment?

Of course, when faced with an insurmountable challenge to a foundational assumption or identity-constituting paradigm, the mind will work incessantly until it has a suitable replacement. Minds are never empty. Some new threads may be barely noticeable in the fabric of existence, but others may be so transformative that they would literally render the entire world unrecognizable. The question then becomes, how can we begin to dismantle the unsustainability mind, knowing it must be quickly replaced with something else, if we do not yet have a guide to lead us out of our zone of proximal development? The problem is that when it comes to sustainability there are no adults in the Cartesian-constructed room to provide parental guidance. This is where the environment-as-parent cultural tool may become most useful.

The dilemma of the unavailability of a fully formed replacement paradigm has led some to argue that environmentalism should be promoted as a religion (Dunlap, 2004). Indeed, Bateson and Bateson (1987) have argued that the solution needs to be formulated as something "like religion" but with caveats, observing, for example, "We have to have in mind not an orthodoxy but a wide and compassionate recognition of the storm of ideas in which we all are living and in which we must make our nests—find spiritual rest—as best we can" (pp. 178–179). Others have looked to aesthetics for hope (e.g., Berleant, 1992, 2005, 2014). I think these efforts are on the right track in rejecting the notion that solutions will come in the form of a proof. They will always, I believe, rely heavily on intuitive best guesses, poetry (Stefanovic, 2000, p. 150), and more holistic ways of knowing than Western science can currently offer. There are, however, forms of guidance available to point us in the right direction, and with practice, traces of potential sustainability *can* be uncovered.

The Traces of Sustainability

Traces of sustainability might emerge from two directions. Either we can imagine, and then learn how to experience, the world through a new sustainability mind, or the world can nudge us toward a new sustainability mind. In a best-case scenario, and certainly the only workable scenario over the longer term, these two actions would occur simultaneously, but each approach is challenging.

The challenge of experiencing the world through a sustainability mind is very much like trying to imagine how prehistoric peoples experienced their Neolithic material-engagement processes, from just the marks, drawings, or stone cutting tools they left to posterity. Not that the prehistoric mind could serve as a model for sustainability, but the gap between our current situation and a sustainability mind is probably as large as that between ourselves and the prehistoric mind of cave paintings and stone axes (Malafouris, 2013a, pp. 179–205).

There are, however, some clues to help imagine how important different mediational means are in constituting different experiences and different minds. We might consider extreme examples where certain aspects of these mediational means are completely lacking, for example, the subjects of King Frederick's experiments or Romanian orphans under Ceaușescu's rule. King Frederick (1194–1250 CE) devised an experiment to discover what language was spoken in the Garden of Eden. This involved separating children from their parents, never speaking to them, and waiting to see what language they would eventually speak. In the twentieth century, Romanian orphanages replicated very similar conditions. In both cases, the children were provided with adequate food and shelter but lacked access to a digestible meaning system compatible with human life. Of course, children can, and often do, survive even extremely bad parenting. What was missing here, though, was the possibility of being woven into the world with enough empathy and *connectivity* to make life *sustainable*.

Yet more extreme examples can be found, where societies literally sacrifice children—a practice that is not as rare as one would hope (Rundin, 2004), and perhaps the worst form of parenting imaginable—although it is surely obvious here that responsibility and blame accrues to the whole of society, not just individual parents. The child sacrifices that took place in the ancient Inca Empire (Andrushko et al., 2011), for example, were surely enabled by serious flaws in the sociocultural context and its mediational means. There had to be an elaborate material culture supporting a ritualistic practice that enabled the ontological shift from children as valuable, sacred, and living constituters and future progenitors of society to an instrumental and sacrificial means for placating deities. More recent examples of child soldiers and bomb carriers may have their roots in the experience of colonialism and/or religious fundamentalism, but they too require a similarly destructive cultural means that ensures that the child is instrumentalized in the extreme. The child becomes a tool for the ritualistic achievement of something the society deems to be more important than sustaining the child's life. But what might we say about a society that, in the name of free markets and individual profit, removed the ecological substratum of the world necessary for our children, and for later generations, to even survive, let alone become sustainable? Can we really not try to imagine something better than this? What we know but "cannot tell", to borrow Polanyi's (1966, p. 8) description of tacit knowledge, *can* hurt us. But examples like these provide instances of environments and mediational means that speak to us about the failure to provide the type of parenting that enables sustainability. They show quite plainly that environments matter, in terms of the affordances they do, or do not, can or cannot, offer.

My children have, in some sense, taught me how to be a parent. They have provided me with the cultural/cognitive tools I use to constitute myself as a parent. I could not have become the parenting-person I am without them to provide me with those navigational skills. This reflexive aspect of the relationship is why parenting is such a useful guiding metaphor for understanding sustainability. What it means to be a "good enough" parent in a world of plastic toys, genetically modified tomatoes, shopping malls, and nuclear bombs remains something of an open question. But it is surely clear that good-enough parenting, just like a sustainable society, cannot be based on the domination or destruction of one side of the equation with the delusional goal of benefiting the other. Sustainability, similarly, implies a balancing partnership between peoples, mediational means, and ecological systems, not mastery or (experimental and/or profiteering) sacrifice of the vulnerable party.

The methodological challenge is to overcome the fact that our tool-constituting and constituted species is highly susceptible at times to both hallucinatory tools and tools that work to destroy the tool user. We are especially susceptible when these tools are inherited and are indexical of environments that have been induced to provide bad parenting. The quandary is that any new tool or knowledge must, as already argued, resonate to some degree within the existing cultural matrix if it is to be adopted. Indeed, it must resonate enough to lead individuals, organizations, and societies out of their ZPD and guide social change in the direction of sustainability. Sustainability, then, seems to involve a process of untethering ourselves from the unsustainability world we have become, dismantling our old self and our old world, to be reborn and reformed into an unknown and unknowable future.

Sustainability can only emerge in "between," as the aggregate consequence of all those social practices. Tsoukas (2005) identifies three dimensions of social practices that might, at least, give us somewhere to think about as a starting point:

> First, the social position or *role dimension*; namely, the normative expectations that are associated with the carrying out of a particular role.
>
> Second, the *dispositional dimension*; namely, the system of mental patterns of perception, appreciation, and action which has been acquired by an individual via past socializations and is brought to bear on a particular situation of action.
>
> Finally, the *interactive-situational dimension*; namely, the specific context of a social activity within which normative expectations and the *habitus* are activated. (p. 104; italics in original)

These dimensions, combined with some knowledge of the characteristics of the given society's ZPD, provide some guidance about where our methods for sustainability might focus. Sustainability implies more than just knowledge. It entails the production and reproduction of patterns of sustainable social practice throughout all three dimensions. At this point, we may not be able to say exactly what such practices would look, smell, sound, or feel like. We can only uncover their traces by uncovering and considering those current patterns of social practice we know to be antithetical to sustainability, patterns like those described by Bateson (1972/2000):

> Ecology, in the widest sense, turns out to be the study of the interaction and survival of ideas and programs (*i.e.*, differences, complexes of differences, etc.) in circuits.
>
> Let us now consider what happens when you make the epistemological error of choosing the wrong unit: you end up with the species versus the other species around it or versus the environment in which it operates. Man against nature. You end up, in fact, with Kaneohe Bay polluted, Lake Erie a slimy green mess, and "Let's build bigger atom bombs to kill off our next-door neighbours." (pp. 490–491)

If our current tools lead us to "make the epistemological error of choosing the wrong unit," then sustainability, within such a context, would seem to be a tall order indeed. A first step, again, might be to simply ask ourselves, as Bateson did, exactly how our efforts at eradicating and mastering nature have been going. Winston (1997) provides a good starting point for our reality check:

> Our war against pests has exacted a heavy toll, but it is a toll we do not see. We have come to expect a sanitized world, a world without cockroaches in our kitchens, insects in our lettuce, dandelions in our lawns, or blemishes on our oranges. The

> sanitized existence has a price, however, in thousands of serious or fatal pesticide poisoning cases each year, especially in developing countries, and in fields, meadows, and forests stripped of biological diversity. Not only do we annihilate pests, but we also take out plants, insects, fish, birds, and myriad other organisms when we spray. (p. 186)

Actually, we do not often succeed in *annihilating* pests. Rather, the dominant scenario runs something like the following: 1) A monoculture (fitting the exigencies and parameters of the civilized/mastered conceptual category) provides a large concentrated food source attracting and enabling a population explosion of the "pest." 2) Rather than question the rightness of the monoculture logic (which would incidentally also call the implicit civilized-untamed split into question), solutions are formulated to eradicate the contagion that the monoculture had previously called forth. 3) Especially following the industrial revolution, when "improved" transportation and the colonial export of a set of wholesale mediational tools became commonplace, this initially localized problem becomes global (Winston, 1997, p. 7). 4). Science and technology are applied, especially following WW II, when methods originally designed to kill people were adapted to the task of killing pests. 5) The chemicals used to exterminate pests have further ecological ramifications, often including killing the natural predators of the targeted pests due to bioaccumulation as the target species is eaten by them. This gives the appearance of a correlation between the continued application of chemicals and a reduction of the pests. 6) Without populations of natural predators, pest numbers increase when pesticide application is discontinued, leading to a kind of addiction to a poison that, if our model is correct, is by no means confined to the physical or material realms but also spreads into the cognitive and affective frameworks of the social mind. This process brings us back to step 1 and is then repeated indefinitely.

Now this means that the ecological effects of our actions provide, *quite literally*—that is not just metaphorically—the conditions of our state of mind and vice versa. MET regards the material world as integral to mind. At a fundamental level, we inhabit a monoculture focused on pest eradication, the erasure rather than cultivation of alternative modes of being, and of a war on nature. Recall, a tragedy of the commons can only occur in a world with "no alternative." The world, as we configure it, comes to reflect the underlying logos and mythos of these communities of practice, whether they are configured into altars for the sacrificing of children, or nature parks, skyscrapers, slums, and toxic waste dumps, or the television sets, automobiles, lawn mowers, and other paraphernalia supporting and being supported by that configurative genealogy. We are like fish unaware of how the lake we swim in directs and constitutes our behavioral and physical development. These tool sets and practices only become matters of conscious concern, rather than tacit knowledge, when they cease to function.

So where is our "data" that this theory of (un-) sustainability might put to use? Where are the traces? Wertsch's (1998) sociocultural analysis, following Burke (1984), suggests a conceptual framework that focuses on five domains: "Act, Scene, Agent, Agency, and Purpose" (p. 13). These five domains are far from easy to bear in mind simultaneously while also considering their relation to (un-) sustainability. In fact, Wertsch states that

> it often seems that we have reached the limits of human understanding when we try to coordinate two or more pentadic elements in a single account of action and motive. For example, some accounts of action that begin with an agent might attempt to incorporate information about the scene as well, but when it comes to extending the account even further by addressing, say how the purpose or instruments used play a role, the picture gets impossibly complex. (p. 16)

I would also argue that these categories, or units of analysis, are unstable; from one moment to the next the scene may become the agent or the purpose may become the act. Nevertheless, they provide a starting point that we can adapt to provide some sampling categories for a methodology of (un-) sustainability. I suggest that we might begin mapping out the traces of (un-) sustainability by turning to the following conceptual categories that, like Wertsch's (1998) pentadic elements, are to be understood as "tools for interpretation rather than reflections of reality" (p. 14). These categories also recognize, as MET suggests, that "most of our thinking takes place outside our heads" (Malafouris, 2015, p. 356) and that "cognition is not a 'within' property; it is an 'in-between' process" (p. 361). These categories are not generally fixed in time and space; they do not exhaust the number of potential categories that could be sampled and certainly are not mutually exclusive, but they have been selected for their accessibility and because they serve well to illustrate the issue of (un-) sustainability. Substantial research literatures also already exist on each of these categories—research that that can be readily brought into the MET framework. The categories selected are these: bodies, tools, places, processes, agency, memory, and atmosphere.

Bodies

There is an extensive literature, for example, from feminist and social theorists, on the many ways in which bodies become indexical of an ecological-sociocultural context (Bordo, 1993; Turner, 1996; Williams & Bendelow, 1998). Bodies are always situated and enculturated bodies, which are inscribed and express aspects of their time and place. Bodies respire, leak, absorb, transmit, transform, align, inure, master, and are mastered, but perhaps more importantly, they provide affordances. In this sense, they overlap with the category of tools, and from a neuroscience and MET perspective; when we work with material tools, the brain treats those tools as coextensive with the body. Bodies themselves, with or without tools, are our intimate mediators of our becoming with the world. Bodies are more *with* than *in* the world. Bodies become in action (or inaction) folding us into and melding with environment. The world also materially infiltrates and constitutes the body. For example, much of the living part of who we are is not even human, but is composed of bacteria on our skin and in our guts.

There is no (un-) sustainability without the complicity of bodies. Importantly, bodies are where things feel right or wrong, and feelings have as much to do with the world and involvement in a situated community of practice as with the body. That being *with* the world can sometimes become suddenly accessible in the felt affects of the last post played at a remembrance service, in an intimate moment with a lover, or through the pain of chemically induced cancer. Suddenly, at such times, our bodies remind us that the Cartesian maps of separateness that organize our thoughts do not actually correspond very well to the territories they purport to represent.

Bodies are also communicative. We tell stories with our bodies but they are also part of a larger story, as Douglas (1975) reminds us:

> The body, as a vehicle of communication, is misunderstood if it is treated as a signal box, a static framework emitting and receiving strictly coded messages. The body communicates information for and from the social system in which it is a part. It should be seen as mediating the social situation in at least three ways. It is itself the field in which a feedback interaction takes place. It is itself available to be given as the proper tender for some of the exchanges which constitute the social situation. And further, it mediates the social structure by itself becoming its image. (p. 83)

MET, of course, emphasizes that bodies are situated and that positionality matters:

> It means, in other words, that the situation (environmental, technological, cultural, or social) can shape and/or become part of the embodied thinking process. By the same token, "being embodied" means that "embodiment" *matters*. It means, in other words, that the body shapes and/or becomes part of the process. (Malafouris 2013a, pp. 221–222)

Bodies, then are always implicitly woven into, and complicit with, any regime, making them an ideal trace of (un-) sustainability.

Tools

Tools situate, extend, distribute, and mediate minds. We master tools but tools also master us. Western minds often think that the key to sustainability is simply better tools, in the sense of novel technics. But, as already argued, tools need to be understood very differently. As our bodies begin to include micro-plastics with serious health issues and as plastics decompose to produce molecules that, for example, mimic the hormone estrogen, we might become aware of a tool called the *precautionary principle*. If only, we might think, we had used the precautionary-principle tool, we might have had a very different outcome, one much more sustainable. But, of course, the precautionary-principle tool has been available for quite some time, and its failure to resonate, and the failure to employ it, points to both the unsustainable constitution of current society and the limits of our current social mind. It also exemplifies the fact that we cannot simply make an unmediated (or un-tooled) decision to adopt a technology. As Malafouris (2013) points out,

> When it comes to tool making and tool using, it is not appropriate to see the brain as the executive controller for embodied activity; rather, it is the other way around: Now embodied activity controls the brain. What we call "knapping" (and it remains for us to define the precise ontological and epistemic contours of that term) is the chief element or the gravitational center of the complex sequences of causes and events that unfold. (pp. 174–175)

It is not just knapped stone tools, but indeed all sorts of tools, including the "precautionary principle," that operate in this way. If we want to meaningfully discuss (un-) sustainability, we need to make a distinction between tools-in-use (tools that resonate) and hypothetical tools (tools that *might* be available one day within a different constellation of human action), and then we need to consider not only the affordances that a tool provides but also the implicit and explicit foreclosures the tool entails. No discussion of (un-) sustainability can be complete without consideration of tools and potential tools.

Place

Place is obviously a contested term, with battles fought over its "true" meaning in geography and philosophy departments around the world (see, for example, Casey, 1997; Cresswell, 2004; Massey, 2005). This, however, is indicative of its importance in many fields from economics (Harvey, 2001) to ethics (Smith, 2001). The fluidity and context dependence of the term suits our current purposes. To understand place, we might return to our metaphor of the parent. What is a parent? Well, it depends. The parent is part of a relationship, but where the parent ends and child begins, who can say? The parent can be an entire world for an infant, or recede to a faint memory for a 20-year-old. Place, like a parent, might be a collection of formative experiences (Tuan, 1974) or where wisdom sits (Basso, 1996). Most importantly, perhaps, we never stop learning about and becoming better acquainted with places or parents. From a sociocultural-analysis perspective, place translates roughly to Wertsch and Burke's "scene" of action which also notes its fluid, almost ephemeral nature.

> When analyzing scenes, Burke made it clear that they do not have some kind of independent, static existence and simply wait to be accurately or inaccurately identified. Instead, like other elements of the pentad, the scene is a tool for interpreting human action and motives … what counts as the scene when making one interpretation of an action may not count as the scene when making another. When we argue over why someone did something by arguing over whether it was just the immediate situation that person was in or whether it was, say, the whole political situation in the United States today, we are arguing over what the relevant circumference of a scene is. Such discussions make it clear that there is often no simple, objective, a priori definition of the scene and its boundaries, a point that does not deter us from relying heavily on the notion of scene when interpreting action. (Wertsch, 1998, pp. 14–15)

We should not then conceptualize place (or scene) as a static background upon which action takes place, as Jones and Cloke (2008) illustrate:

> Places are recognized as temporal processes where all manner of trajectories—of people, non-humans, economies, technologies, ideas and more—come, are brought or are thrown together to assemble enduring, but also changing, formations which settle out into distinctive patterns of places, yet which are still fully networked into the wider world. (p. 87)

Place, in other words, operates in mind as well as occurring in the world. From the MET perspective, place becomes more like a full partner within the mediated action of (un-) sustainability. There is no unmediated experience of place. Even the nowhere of screen technology relies on the education of visual sensory systems by the parenting-like influences of environment.

Place acts like a parent in providing, or failing to provide, a nurturing environment. Place educates the senses, teaching us what to pay attention to and what to disregard. Place holds parts of our distributed and extended minds in trust for us. Our thoughts and emotions can never completely escape from place just as our parents always leave an indelible mark on our minds. Sometimes the most mundane and ordinary place can still afford the degree of comfort required for our thoughts to wander to faraway places. Unfamiliar environs may hold our attention and keep our endocrine systems at the ready for any unexpected threat or emergency. The traces of (un-) sustainability always pass through, are always accessed through, places.

Narrative

Narrative is another element of Wertsch's sociocultural-analysis framework and another way of holding human action together, organizing and structuring it, giving it texture and purpose. It is also, of course, a mediational means or cultural tool, like metaphor, frame, language, or paradigm, and one that provides a certain coherence to the ways those other tools hang together through time and in places. Narrative is not simply an interpretive tool, but rather a dominant characteristic of the structure of meaning. Narratives do not necessarily require spoken or written language but, from the perspective of MET, may also be inscribed in material culture or geographic configurations. Whenever things become meaningful through participation in action there is a narrative, and patterns of sensibilities are often experienced and organized schematically into the form of narratives. There are, in other words, tacit narratives that parallel tacit knowledges. Narratives, whether tacit or explicit, form a zigzag relationship with purpose. There is no purpose without an enabling narrative and no narrative without at least an implicit purpose. That is why narrative is critical to any discussion of (un-) sustainability, but it is equally important to remember that the agency exercised by narrative is generally part of a distributed agency. Making changes only within the domain of a specific narrative provides no guarantee of broader social-ecological change across other domains. Morton (2013) illustrates this point well:

> Most environmental art is constructivist, not object oriented, for the simple reason that constructivism has been the dominant mode of art in modernity. Another strange fact: the art that emerges at the inception of the Anthropocene is not fully capable of addressing it. The 350.org movement places the number 350 on gigantic surfaces, such as beaches, that are photographed from above. Like the Nazca Lines in Peru, they cannot be seen fully from ground level. Three hundred and fifty is the number of parts per million CO^2 that is deemed acceptable to avoid disastrous global warming. This graffiti is designed to change your mind. (p. 188)

Morton points out the inherent limitation to the power of narrative, even when the narrative is inscribed onto the Earth's surfaces. We might add that in a screen culture, the agency of photographs is certain to evaporate quickly. For broad-based social change to take root, the narrative needs to be widely distributed and given a certain permanence and/or reproducibility in landscape and architecture as much as discourse and rhetoric.

Again, the parenting metaphor comes to mind. Parents introduce and habituate children to certain forms and genres of story. Expectations are established. Cues are instinctively sought. The mode of retelling, the cadence, the feel of paper and book covers, the family circumstances, the places invoked and the present place all implicitly inform the meanings taken from the story. We should be wary of the idea of a narrative that by itself has the power to "change your mind." That story belongs to the Cartesian delusion where thought is believed capable of exercising sovereignty over the material world.

Process

Process is what forms recognizable patterns in the world. Process is about the pattern of action that can be seen when bodies, tools, and/or places or their component parts intersect. We might think of plate tectonics and continental drift organizing landmasses into recognizable geomorphological patterns, or meteorological processes organizing the atmospheric and hydrological cycles into recognizable patterns of weather and climate. According to Capra (1996),

> There is a ceaseless flux of matter through a living organism. Each cell continually synthesizes and dissolves structures and eliminates waste products. Tissues and organs replace their cells in continual cycles. There is growth, development, and evolution. Thus from the very beginning of biology, the understanding of living structure has

> been inseparable from the understanding of metabolic and developmental processes. (pp. 159–160)

There is a continual dance between form and/or structure and process. Even a strictly cognitive process will involve cogitation about something that leaves an imprint on the neural structure of the cogitator and traces of action in the form of a changed environment. The pattern of how change is etched into the forms and structures of the world is what I will refer to as process. The *pattern* that emerges from the recurrent expressions of agency or action constitutes a process.

Agency

The term *agency* gets at the dominant determinants or parameters of an action or process. Recurrently exercised agency forms into an identifiably path dependence and that, in turn, is what constitutes a process. More random instances of action need to be dealt with as instances of agency, but there are also always many constituent instances of agency within any discernable process. We can think of the process of people growing rice, for example. Within that process, there are mediated actions including the agencies exercised by the people, their stories, the land, the rice, etc. On the other hand, we would encounter real difficulty in identifying the specific agencies associated with the evolutionary processes that etch changes into the original land and ecosystem, including the integration of humans into that broader unfolding.

The exercise of agency is always embedded within broader coextensive processes. The main problem of the Cartesian perspective arises because we insist on modeling agency as a matter of human thought, as that ghost in the machine, identified as the (sole) instigator of action (Latour, 1993, 2005). But even here, as Gallagher and Rasom (2016) explain, "people don't act outside of some context or simply from some *de novo* intention formed abstractly inside their mind" (p. 346). Likewise, for things to exert agency, they must be embedded within a context of mediated action. The exercise of agency is always a co-constituting material engagement *with* the world.

Memory

Remembering can occur in brains, but it can also occur in cultural tools such as stories, myths, metaphors, material artifacts, in bodies, places, or whenever and wherever traces remain and continue tracing. Memory is generally focused through some type of mediational means. Memory can act to slow or regulate some processes. Memory can be co-constituent within action. Memory has been an important tool for spreading the Western-Cartesian unsustainability. This passage by Matunga (2013) is illustrative:

> In New Zealand, as in most countries colonized by settler governments such as Australia, Canada, and the United States, the materiality (i.e., physical quality, presence, and structure) and memory (i.e., recall of experience, even existence) of Indigenous communities has generally been erased. In the cities it was replaced with imperial monuments, colonial buildings, colonial cathedrals, colonial gardens, and colonial city patterns modelled on the "old world" and the mother country. In the New Zealand context, many cities were "designed" in England and simply migrated with the settlers and dropped on the landscape to create a new home away from home. The countryside saw Indigenous biota cleared, native land carved up and replaced with farm lots for the new settlers. The aim was to remove any material evidence/reminder and memory of Indigenous communities, their places, sites, resources, and villages, and replace it with a new colonial order, ultimately creating a "new" materiality and memory for/of settler communities. (pp. 8–9)

Memory, in other words, is about what persists, but also what is negated through erasure, whether in bodies, tools, narratives, or places. For memory to be associated with sustainability implies skill, including both skillful remembering and skillful forgetting.

Atmosphere

Like place, atmosphere is as much about mind as it is about world. Atmosphere is constituted by all the other elements of (un-) sustainability as they are experienced by the body/brain. Atmosphere is generally referred to as the subjective experience of a place, but actually that place clearly plays all kinds of roles in producing that atmosphere, as in the atmosphere of the crowded sports stadium, the forest path, or the busy street. Atmosphere can be manipulated, for example, according to the precepts of the marketing model. As Quartier, Vanrie, and Van Cleempoel (2014) explain: "In many cases the goal of creating a certain atmosphere in a retail environment is precisely to try to elicit a certain emotional effect in the consumer in order to achieve some desired behavioral result" (p. 33). In this sense, atmosphere might be the product of a staged capitalist environment, of a profit motive, but even this relies on particular models of subjectivity, aesthetics, and material possibilities (the *tools*) available to manipulate experiences. Some atmospheres have the power (or agency) to overwhelm across many cultural models of subjectivity or sensibilities, both individual and collective; othersare much more specific. It is this capacity, the potential to connect sustainable with unsustainable models of subjectivity that makes atmosphere especially interesting as an index of (un-) sustainability.

These eight parameters of (un-) sustainability provide my units of analysis and my samples as I outline the traces of (un-) sustainability. Any hope for sustainability will depend on the capacity for cascading changes across and between these eight (as well as other) dimensions, changes that can overcome the inertia in other dimensions.

References

Abel, C. (2015). *The extended self: Architecture, memes, and minds*. Manchester, UK: Manchester University Press.

Andrushko, V. A., Buzon, M. R., Gibaja, A. M., McEwan, G. F., Simonetti, A., & Creaser, R. A. (2011). Investigating a child sacrifice event from the Inca heartland. *Journal of Archaeological Science*, 38, 323–333.

Appadurai, A. (1986). *The social life of things: Commodities in cultural perspective*. Cambridge, UK: Cambridge University Press.

Barry, J. (2012). *The politics of actually existing unsustainability: Human flourishing in climate-changed, carbon-constrained world*. Oxford, UK: Oxford University Press.

Basso, K. H. (1996). *Wisdom sits in places*. Albuquerque: University of New Mexico Press.

Bateson, G. ([1973] 2000). *Steps to an ecology of mind*. Chicago, IL: University of Chicago Press. (Original work published 1972).

Bateson, G. (2002). *Mind and nature: A necessary unity*. Cresskill, NJ: Hampton Press.

Bateson, G., & Bateson, M. C. (1987). *Angels fear: Towards an epistemology of the sacred*. New York, NY: MacMillan.

Bazerman, M. (2006). Climate change as predictable surprise. *Climate Change*, *77*(1–2), 179–193.

Bean, J., Figueiredo, B., & Larsen, H. P. (2017). The object of research: Considering material engagement theory and ethnographic method. *Ethnographic Praxis in Industry Conference Proceedings*, *2017* (1), 262–272.

Bennett, J. (2004). The Force of Things. Political Theory, 32(3), 347–372.

Bennett, J. (2010). *Vibrant matter: A political ecology of things*. Durham, NC: Duke University Press.

Berkes, F., Colding, J., & Folke, C. (2003). *Navigating social-ecological systems: Building resilience for complexity and change*. Cambridge, UK: Cambridge University Press.

Berkes, F., & Folke, C. (1998). *Linking social and ecological systems: Management practices and social mechanisms for building resilience*. Cambridge, UK: Cambridge University Press.

Berkes, F., & Ross, H. (2016). Panarchy and community resilience: Sustainability science and policy implications. *Environmental Science & Policy*, *61*, 185–193.

Berleant, A. (1992). *The aesthetics of environment*. Philadelphia, PA: Temple University Press.

Berleant, A. (1994). Beyond disinterestedness. *British Journal of Aesthetics*, *34*, 242–255.

Berleant, A. (2005). *Aesthetics and environment: Variations on a theme*. Aldershot, UK: Ashgate.

Berleant, A. (2014). Environmental sensibility. *Studia Phaenomenologica*, *14*, 17–23.

Boivin, N. (2008). *Material culture, material minds: The impact of things on human thought, society, and evolution*. New York, NY: Cambridge University Press.

Bordo, (1993). *Unbearable weight: Feminism, Western culture, and the body*. Berkeley: University of California Press.

Boulding, K. (1978). *Ecodynamics: A new theory of societal evolution*. Beverley Hills, CA: SAGE.

Bourdieu, P. (1977). *Outline of a theory of practice*. Cambridge, UK: Cambridge University Press.

Bruner, J. (1986). *Actual minds, possible worlds*. Cambridge, MA: Harvard University Press.

Burke, K. (1966). *Language as symbolic action: Essays on life, literature, and method*. Berkley: University of California Press.

Burke, K. (1984). *Permanence and change: An anatomy of purpose* (3rd ed.). Berkeley: University of California Press.

Bushell, B., & Goto, M. (2011). Leadership for a sustainable society: A transformative learning approach. *Procedia—Social and Behavioral Sciences*, *29*, 1244–1250.

Calleja, C. (2014). Jack Mezirow's conceptualization of adult transformative learning. *Journal of Adult and Continuing Education*, *20*, 117–136.

Candlin, F., & Guins, R. (2009). *The object reader*. London, UK and New York, NY: Routledge.

Capra, F. (1996). *The web of life: A new understanding of living systems*. New York, NY: Anchor Books.

Casey, E. S. (1997). *The fate of place: A philosophical history*. Berkeley: University of California Press.

Clark, A. (1997). *Being there: Putting brain, body, and world together again*. Cambridge, MA: MIT Press.

Clark, A. (2003). *Natural-born cyborgs: minds, technologies, and the future of human intelligence*. New York, NY and Oxford, UK: Oxford University Press.

Cleeremans, A., & Jiménez, L. (2002). Implicit learning and consciousness: A graded, dynamic perspective. In R. M. French & A. Cleeremans (Eds.), *Implicit learning and consciousness: An empirical, philosophical and computational consensus in the making*. New York, NY: Psychology Press.

Cochrane, A. (2008). Some stimulating solutions. In C. Knappett & L. Malafouris (Eds.), *Material agency: Towards a non-anthropocentric approach* (pp. 157–186). Toronto, ON: Springer.

Cresswell, T. (2004). *Place: A short introduction*. Oxford: Blackwell.
Diamond, J. (1999). *Guns, germs, and steel: The fates of human societies*. New York, NY: Norton.
Donald, M. (1991). *Origins of the modern mind: Three stages in the evolution of culture and cognition*. Cambridge, MA: Harvard University Press.
Donaldson, D., & Macpherson, F. (2017, July). Müller-Lyer illusion. In F. Macpherson (Ed.), *The Illusions Index*. Retrieved from https://www.illusionsindex.org/ir/mueller-lyer
Douglas, M. (1975). *Implicit meanings: Essays in anthropology*. London, UK: Routledge & Kegan Paul.
Dunlap, T. R. (2004). *Faith in nature: Environmentalism as religious quest*. Seattle: University of Washington Press.
Farley, H. M., & Smith, Z. A. (2014). *Sustainability: If it's everything, is it nothing?* London, UK: Routledge.
Ferguson, J. (2009). The uses of neoliberalsim. *Antipode, 41*(s1), 166–174.
Fry, T. (2012). *Becoming human by design*. London, UK: Berg.
Galbraith, J. (1958/1998). *The affluent society*. Boston, MA: Houghton & Mifflin.
Gallagher, S., & Ransom, T. G. (2016). Artifacting minds: Material engagement theory and joint action. In G. Etzelmüller & C. Tewes (Eds.), *Embodiment in evolution and culture* (pp. 337–352). Tübingen, DE: Mohr Siebeck.
Gell, A. (1998). *Art and agency: An anthropological theory*. Oxford, UK: Clarendon Press.
Gosden, C. (1994). *Social being and time*. Oxford, UK: Blackwell.
Gosden, C. (2005). What do objects want? *Journal of Archaeological Method and Theory, 12*, 193–210.
Hardin, G. (1968). The tragedy of the commons. *Science, 162*, 1243–1248.
Harries-Jones, P. (1995). *A recursive vision: Ecological understanding and Gregory Bateson*. Toronto, ON: University of Toronto Press.
Harrison, R. P. (1992). *Forests: The shadow of civilization*. Chicago, IL: University of Chicago Press.
Harvey, D. (2001). *Justice, nature & the geography of difference*. Malden, MA: Blackwell.
Hodder, I. (2012). *Entangled: An archaeology of the relationships between humans and things*. Chichester, UK: Wiley-Blackwell.
Holling, C. S., Berkes, F., & Folke, C.(1998). Science, sustainability and resource management. In F. Berkes & C. Folke, (Eds.), *Linking social and ecological systems: Management practices and social mechanisms for building resilience* (pp. 342–362). Cambridge, UK: Cambridge University Press.
Hundert, E. M. (1997). *Lessons from an optical illusion: On nature and nurture, knowledge and values*. Cambridge, MA: Harvard University Press.
Hutchins, (1995). *Cognition in the wild*. Cambridge, MA: MIT Press.
Ingold, T. (2000). *The perception of the environment: Essays in livelihood, dwelling and skill*. London, UK: Routledge.
Ingold, T. (2013). *Making: Anthropology, archaeology, art and architecture*. London, UK: Routledge.
Jones, O., Cloke, P. (2008). Non-Human Agencies: Trees in Place and Time. In: Knappett C., Malafouris L. (eds) Material Agency. Springer, Boston, MA.

Kim, H., Yun, Y., Ryu, S., Chang, Y.,Kwon, M., Cho, J., & … Kim, H. (2018). Correlation between gut microbiota and personality in adults: A cross-sectional study. *Brain, Behavior, and Immunity*, 69, 374–385.

Knappett, C. (2005). *Thinking through material culture: An interdisciplinary perspective*. Philadelphia: University of Pennsylvania Press.

Lakoff, G., & Johnson, M. (1980). *Metaphors we live by*. Chicago, IL: University of Chicago Press.

Latour, B. (1993). *We have never been modern*. Cambridge, MA: Harvard university Press.

Latour, B. (2005). *Reassembling the social: An introduction to actor-network-theory*. Oxford, UK: Oxford University Press.

Lozano, R., Ceulemans, K.,Alonso-Almeida, M., Huisingh, D., Lozano, F. J., Waas, T., Hugé, J. (2015). A review of commitment and implementation of sustainable development in higher education results from a worldwide survey. *Journal of Cleaner Production*, *108*, 1–18.

Malafouris, L. (2004). The cognitive basis of material engagement: Where brain, body and culture conflate. In E. DeMarrais, C. Gosden, & C. Renfrew (Eds.), *Rethinking materiality: The engagement of mind with the material world* (pp. 53–62). Cambridge, UK: MacDonald Institute for Archaeological Research.

Malafouris, L. (2008). At the potter's wheel: An argument for material agency. In C. Knappett & L. Malafouris (Eds.), *Material agency : Towards and non-anthropocentric approach* (pp. 19–36). New York, NY: Springer.

Malafouris, L. (2013a). *How things shape the mind: A theory of material engagement*. Cambridge, MA: MIT Press.

Malafouris, L. (2013b). Mind into matter. *New Scientist*, *219*(2933), 28–29.

Malafouris, L. (2015). Metaplasticity and the primacy of material engagement. *Time and Mind*, 8(4), 351–371.

Malafouris, L., Gosden, C., & Overmann, K. A. (Eds.). (2014). Creativity, cognition and material culture. *Pragmatics & Cognition*, *22*(1), 1–4.

Malewski, E., & Jaramillo, N. (2011). *Epistemologies of ignorance in education*. Charlotte, NC: Information Age.

Massey, D. (2005). *For space*. London, UK: SAGE.

Matunga, H. (2013). Theorizing indigenous planning. In R. Walker, T. Jojola, & D. Natcher (Eds.), *Reclaiming indigenous planning*, (pp. 3–34). Montreal, Kingston: McGill-Queen's Press.

Mezirow, J. (2000). *Learning as transformation: Critical perspectives on a theory in progress*. San Francisco, CA: Jossey-Bass.

Morton, T. (2013). *Dark ecology: For a logic of future coexistence*. New York, NY: Columbia University Press.

Popkewitz, T. (2017). The promise of empirical evidence and benchmarks: The Lorelei's whispers. In T. Salonen & H. Lindquist (Eds.), *Knowledge for change lecture series 2*. Malmö, SE: Malmö University. Retrieved from https://blogg.mah.se/knowledgeforchange/files/2017/12/Thomas-Popkewitz-article-print-version.pdf

Quartier, K., Vanrie, J. and Van Cleempoel, K. (2014). As real as it gets: What role does lighting have on consumer's perception of atmosphere, emotions and behaviour? *Journal of Environmental Psychology*, 39, 32–39.

Renfrew, C. (2001). Symbol before concept: Material engagement and the early development of society. In I. Hodder (Ed.), *Archaeological theory today* (pp. 122–140). Cambridge, UK: Polity.

Renfrew, C. (2003). *Figuring it out: What are we? Where do we come from? The parallel visions of artists and archaeologists*. New York, NY and London, UK: Thames & Hudson.

Robbins, T. (2007). *Lawn people: How grasses, weeds, and chemicals make us who we are*. Philadelphia, PA: Temple University Press.

Rundin, J. S. (2004). Pozo Moro, child sacrifice, and Greek legendary tradition. *Journal of Biblical Literature, 12*, 425–447.

Sandri, O. J. (2013). Threshold concepts, systems and learning for sustainability. *Environmental Education Research, 19*, 810–822.

Schumpeter, J. A. (2005). *The theory of economic development: An inquiry into profits, capital, credit, interest, and the business cycle* (11th ed.). New Brunswick, NJ: Transaction. (Original work published 1934).

Searle, J. R. (1983). *Intentionality: An essay in the philosophy of mind*. Cambridge, UK and New York, NY: Cambridge University Press.

Smith, M. (2001). *An ethics of place: Radical ecology, postmodernity, and social theory*. Albany, NY: SUNY Press.

Stefanovic, I. (2000). *Safeguarding our common future: Rethinking sustainable development*. Albany, NY: SUNY Press.

Sweeny, L. B. (2017). All systems go! Developing a generation of "Systems-Smart" kids. In E. Assadourian (Ed.), *EarthEd: Rethinking education for a changing planet* (p. 143). Washington, DC: WorldWatch Institute.

Theiner, G., & Drain, C. (2017). What's the *matter* with cognition? A "Vygotskian" perspective on material engagement theory. *Phenomenology and the Cognitive Sciences, 16*, 837–862.

Tilley, C. (1991). *Material culture and text: The Art of ambiguity*. New York, NY and London, UK: Routledge.

Tilley, C. (1994). *A phenomenology of language*. New York, NY: Oxford: Berg.

Tilley, C. (1996). *An ethnography of the Neolithic*. Cambridge, UK: Cambridge University Press.

Tilley, C. (1999). *Metaphor and material culture*. Oxford, UK and Malden, MA: Blackwell.

Tsoukas, H. (2005). *Complex knowledge: Studies in organizational epistemology*. New York, NY: Oxford University Press.

Tuan, Y. (1974). *Topophilia: A study of environmental perception, attitudes, and values*. New York, NY: Columbia University Press.

Turner, B. S. (1996). *The body and society*. London, UK: SAGE.

Voltaire. (2007). *Candide, ou l'optimisme, nouv. éd.* Paris, FR: Galimard. (Original work published 1759).

Vygotsky, L. V. (1978). *Mind in society: The development of higher psychological processes*. Cambridge, MA: Harvard University Press.

Vygotsky, L. V. (2012). *Thought and language: Revised and expanded edition*. Cambridge, MA: MIT Press. (Original work published 1962).

Weiss, A. (1995). *Mirrors of infinity: The French formal garden and 17th century metaphysics*. New York, NY: Princeton Architectural Press.

Wertsch, J. V. (1991). *Voices of the mind: A sociocultural approach to mediated action*. Cambridge, MA: Harvard University Press.

Wertsch, J. V. (1998). *Mind as action*. New York, NY: Oxford University Press.

Wheeler, M. (2005). *Reconstructing the cognitive world: The next step*. Cambridge, MA: MIT Press.

Williams,S. J., & Bendelow, G. (1998). *The lived body: Sociological themes, embodied issues*. London, UK: Routledge.

Winston, M. L. (1997). *Nature wars: People vs. pests*. Cambridge, MA: Harvard University Press.

· 3 ·

LET'S GO FISHING!

The hypothetical utterance "let's go fishing" would not be possible without specific instances of an implicit, pre-existing, and recursive materially engaged ecology of mind. Such a process of material engagement would include bodies; tools (both cultural/cognitive and material); pre-existing ideas and memories (distributed among living beings, including fish) and material things (e.g., waters with fish in them); places; stories; processes; agents; and all these together constituting a richly meaningful atmosphere implicit in any invitation from a parent to a child to go fishing. The purpose of this chapter is to flesh out the implicit and recursive dimensions of such a common everyday invitation and develop theoretical connections between those dimensions pertinent to the contemporary societal imperative of sustainability.

To recap, the primary focus of this book is on the materiality of social cognition, putting mind and matter back together, and the implications of already-in-practice matter-and-mind assemblages as they relate to (un-) sustainability. As outlined in the previous chapter, the perspective I adopt begins from an understanding of cognition as "embodied, distributed, mediated, situated, extended and enacted" (Malafouris, 2004, p. 57) within material-engagement processes. From a Material Engagement Theory (MET) perspective, brain, body, tool(s), and object/materiality are co-constitutive of cognitive processes. A mind requires each of these elements to be able to function as a mind. Minds always develop within bodies acting within sociocultural and sociomaterial contexts to the point where attempts at disentanglement become simply exercises in obfuscation. In worse-case scenarios, a focus on the material without attention to the cognitive or vice versa leads to misperception and misunderstanding and the institutionalization and normalization of extremely self-destructive habits. The achievement of sustainability can only result from the orchestration of a sort of harmony, a "tuning" (Benedito, 2017) between all the component parts of embodied, distributed, mediated, situated, extended, and enacted action, including brains, bodies, material objects, narratives, meaning systems, biogeochemical processes, the laws of physics and chemistry, and so on. We might think of this configuration as a partnership mind as opposed to the Cartesian control/controlled mind.

Establishing a sort of mental/emotional cartography of sustainability is a difficult but necessary goal. The resultant maps may seem to represent something as alien to the reader as Galileo's account of sun and Earth, the first glimpse of pond water through a microscope, or the Darwinian concept of natural selection would have initially appeared. The human mind relies heavily on maps, and the maps available tend to become deeply embedded in the synaptic structures of individual and social minds. In individual brains, these structures are physiological, but across societies they involve the various shared tools, both material and conceptual, used to give form and pattern to social cognitive processes. The key variable in such circumstances is the depth to which a map gets sunk into a habitus, forming an important part of the user's identity, whether individual or collective. When a dominant, socially resonant map is suddenly recognized as untrustworthy, the resultant experience of being lost can be emotionally and intellectually disruptive and tends to be resisted vigorously, the force of the experience depends largely on the how accustomed one has become to the maladaptive (or simply erroneous) features of the internalized map (see also Kuhn, 1962/1996).

Generally, the Western-Cartesian perspective induces a heightened skepticism of narrative forms of knowledge that are considered "softer" and less trustworthy than the propositional and quantitative forms of knowledge produced by Western-scientific methods. Yet without attention to narrative forms, we are forced to accept a positivist stance by default. Following Wertsch (1998), I will pay particular attention in this chapter to the role of "narrative as a cultural tool" (pp. 73–108) and what better way to do that with than a discussion of fishing?

When we become certain that a fish is *just* a fish—or perhaps a bundle of genetic information from a scientific perspective, or a commodity found in the frozen food section of the supermarket from the general public's perspective—then the suggestion that a fish is something quite different—that, for example, it is irreplaceably constitutive of a much broader way-of-becoming—challenges well-established notions of self-identity. As our understandings and intuitive definitions of fish change, so do our understandings and definitions of ourselves. What, after all, is a self if not a cumulative history of experience resulting in a set of internalized maps of the world, including maps of fish and fish-place-human assemblages? Fish generally pre-exist us. We are born into a world with fish. We may invent, import, or otherwise produce new maps during our lifetimes, but in general, we begin in childhood with an inherited map to answer the question, What is a fish? Our maps of fish are gleaned from, and redrawn in relation to, our experience in that world with fish. We learn how people, places, and things relate to fish and our cognitive maps of fish become essentially maps of those relations. Fish become things that are eaten at the table, fed in a fish tank, caught from the river, or encountered in bedtime stories.

In the following thought experiment, we will consider the simple act of a parent inviting a child to go fishing. This is not a description of one actual event but rather several imagined, hypothetical events combined with some actual remembered accounts added for illustration. The example of such an invitation is interesting because of the multitude of contingencies necessarily implicated in this once widely practiced activity. Any specific instance of a given parent inviting a given child to go fishing will necessarily be as unique as any specific parent/child/shared meaning system/place assemblage. Such an invitation will always involve a process that is partly based on acquired knowledge and experience and partly on an imagined experience of being and doing the proposed activity (fishing). The following discussion focuses on combinations and permutations of such hypothetical pre-existing and imagined assemblages to illustrate how these assemblages relate to the societal goal of sustainability.

Bodies

What are the domains of materiality implicit in the act of a parent uttering the words, "let's go fishing" to a child and how does such an action relate to the societal goal of sustainability? Leaving aside the question of agency for the moment, the child and the parent obviously exist as physical bodies in a physical and material space. Those bodies provide specific types of affordances and constraints, especially with respect to the proposition, "let's go fishing." Depending on the physical ability of the child, some forms of fishing, such as jigging for Atlantic cod or fishing for great white sharks or large tuna, may be categorically ruled out, possibly changing the nature of the proposition from a sincere invitation to a sarcastic joke. Or perhaps active participation in catching real fish was never sincerely intended in the invitational utterance. Children have an uncanny ability to participate in imagined activities with the most meager of props, such as a stick for fishing pole and a rug for water. It could also be that the proposal was only about being present to watch and learn while the parent fishes. Bodily limitations will proscribe certain types of fishing. Wading in streams requires a certain physical height and strength. Fishing at night requires a body that will not be overcome with fear of the dark. Watching a bobber for movement for hours on end requires a certain ability to concentrate, and so on. Older bodies face other challenges. An older parent may be reluctant to take a difficult path to a previously favorite fishing hole, or to engage in long stretches of trolling in a boat with uncomfortable seats that would push the limits of an arthritic condition. Small-sized hooks may become too much trouble as vision atrophies with age.

Beyond the physical capacities of specific bodies, there are also embodied knowledges or skills that come into play with activities such as fishing. Fly fishing, whether using dry or sinking flies, for example, requires close, felt attention to the fly but also a capacity for un-thought or instinctive reaction to set the hook when the fish strikes. Casting a fly requires a timing that must be felt rather than thought. Trout tickling requires, from my experience, a rare ability to clear one's mind of all thought as the trout always seems to be able to perceive even the slightest whiff of intention in the hand of the tickler. Human bodies are generally not well suited for catching fish without some form of tool, such as net, rod and reel, cognitive maps, or maybe even, in extreme cases, dynamite. In other words, human bodies generally require some extending to be successful in the task of catching fish. Not all bodies are equal in this regard. Small children generally do better when fishing with worms or cheese that the fish will swallow with the hook, rather than requiring the hook to be more skillfully set by the fisher. Some nets require throwing. Some fishing methods require the fish to be seen by the angler. These techniques require varying levels of physical capacity but also varying degrees of embodied skill and/or knowledge.

The achievement of access or the bringing into a "community of practice" of specific bodies represents the normal operation of the parent/child/meaning system/place assemblage. This developmental process is always as much about the building of physical (embodied) capacity as it is about developing cognitive capacity. The timing of flycasting is more about embodied feeling than about theoretical knowledge and can become more difficult if the fly caster tries to concentrate too closely on the process. The learning process is generally greatly facilitated by simply watching an expert. This is another example of an extended self, as the mind of the unexperienced observer begins to operate *as if* it were actively participating in the casting. The expert demonstrates how the fly rod becomes an extension of the fisherperson, providing a template that the initiate can aspire to. This applies as much to the task at hand as it does to the entire experience of fishing. By merely watching the expert, the initiate can deduce the answers to such questions as: What is a fish? and What is fishing? The initiate will also learn the appropriate repertoires of emotional responses to various fishing experiences, such as breaking the line, forgetting the net, falling overboard, or even catching a fish.

Tools

Tools can be material and/or cultural/cognitive. A small child may require little in terms of material tools to engage in activity that would qualify as fishing. Perhaps a pail of water and some plastic toys or a pair of Billy Boots and a small net for exploring along the shore of a pond in shallow water. At the other extreme, tools might include sophisticated radar equipment, expensive boats and motors, rods, reels, nets, lines, sinkers, bobbers, and various other assorted paraphernalia, not to mention lures and baits of every size, color, and shape imaginable. Detailed weather reports may be consulted to determine barometric pressure, precipitation probabilities, or tidal and lunar cycles.

These material tools play a constitutive role in what exactly the act of fishing eventually becomes and the identities of peoples who engage in those fishing processes. We saw above how physical and cognitive aspects of bodies set the parameters of activity and then determine how those bodies, in turn, become constituted as fishers (from initiate to expert) by that process. In like manner, the characteristics of the material tools used in the performance of the activity also determine the nature of the activity and work to co-constitute the nature of the participants or "actants" in that activity (Schmidgen, 2014). Successful fishing generally involves bringing the activity into alignment with the exigencies of the specific waters and species of fish to be fished using quite specific tools. There is always a challenge of increasing the level of skill involved in the various processes, so that, from a Vygotskian perspective, fishing entails the psychological development of the participants. Our identities, in other words, are acquired by the cumulative acts of putting specific tools into specific practices. Cultural/cognitive tools are required for the psychological development of the fisher from initiate to expert. This learning process changes the nature of the learner, the nature of that learner's action in the world, and in consequence, the nature of that world. It is this process of becoming that eventually determines how sustainable any society can be. This (un-) sustainable developmental process operates on both the micro and macro scales at a rate that correlates with the rate of development of new tools, both material and cognitive.

Most children will not be introduced directly to fly-fishing. It is much more likely they will be introduced to fishing as an activity involving worms and bobbers. There will generally be a subtle difference in the level of respect that is learned for the fish depending on the type of fishing that is modeled by the adult. This, of course, can change with time and as the initiate becomes an expert, many mediators will become available for setting the relationship between angler and fish or between angler and angling equipment, for that matter. Fiberglass fly rods became widely available in the 1950s, but split bamboo was still the norm. Today graphite is the standard and split-bamboo rods are antiques worth collecting because there are almost no split-bamboo rod makers left alive. The type of equipment one eventually chooses will influence and be influenced by the image of fishing one becomes acquired by. These images and other tools ultimately establish the identities of individuals and communities.

I have often fished with my uncle's split-bamboo fly rod. During the entirety of those experiences, I have been concerned with not breaking the rod. Although both bamboo and graphite rods are constructed from one continuous piece, or "blank," that is cut to make the sections of the rod, a broken bamboo rod will generally not be reparable. The makers of graphite rods will often try to match the broken section of a rod with another piece of graphite, but the result is a rod made from two different pieces of graphite and the rod will never fish quite the same again. Yet there is a certain reverence for bamboo rods that is missing with graphite. Material objects often can acquire a certain aspect of sacredness when they are used in important events. When I take out a favorite fly rod in winter or tie a certain pattern of trout fly, I am often transported back to the specific waters and the specific trout of past experiences with those tools. Bamboo is in some ways an inferior material for making fly rods, at least from a purely technical perspective, to graphite. Bamboo is heavier and will not take as much line to load on the backcast, resulting in less casting distance, *ceteris paribus*. Yet a bamboo fly rod can have the capacity to connect one to a previous world and mediate the person-tool-place triad in increasingly rare ontological assemblages. Bamboo *connects* to the rapidly vanishing halcyon days of pristine waters, abundant insect hatches, and large trout.

A material tool is not entirely the same thing as a conceptual tool such as a metaphor or vocabulary, or a conceptual image such as a supply-demand model, or a language, or the various subconscious frames we use to make instant suppositions about the character of someone, simply by glancing at their appearance. But although they are not the same, they do operate in some very similar ways. The tool becomes part of the identity of the user and part of the user's way of interacting within environment. As we have already seen, Vygotsky's (1978) classic description of this process involves the use of a knot in a handkerchief as a way of remembering to buy bread. This scaffolding does not always require a material tool but can also operate with a method, such as learning how to do long division (Wertsch, 1998) or how to catch a fish. Learning to catch fish represents the kind of situated learning that accounts for most of the subconscious ways of being and becoming in the world we acquire and, in turn, are acquired by simply by living in a given social-material context of action and community of practice. More than *Homo sapien* (the wise human) or *homo economicus* (rational economic man), we are more rightly self-described and more accurately known as *homo faber*, the tool user (Ferraro & Reid, 2013). Intelligence and wisdom, or lack thereof, lies primarily in the tool and not the user.

Place

There is always an implicit element of place (or scene from a sociocultural-analysis perspective) in any invitation to go fishing, even if it only involves the "nowhere" of a computerized-fishing-game experience. Places also act in some ways like cultural/cognitive tools. Place necessarily informs and is informed by the "entering into" of a material-engagement assemblage. This evolving material engagement also always has sustainability implications. Fishing often involves a deepening engagement and a deepening awareness of the entanglements of place, a deepening interpenetration between place and person—the emergence of an increasingly palpable person-and-place experience as opposed to the Cartesian ghost in a machine located on an abstract tabula rasa. Increasingly, though, in the Anthropocene, fishing, like almost all other human actions, involves letting go of once near-sacred and almost-magical places as anthropogenic processes result in ecological and aesthetic changes to the nature of places to fish. These changes reflect broader environmental trends, and as neoliberal narratives and materializations become increasingly dominant, this process of disenchantment spreads and gathers momentum. The economistic tack of these changes is reflected in the evolving nature of what it means to be human for both individual and social identities.

Consider a place I used to go fishing as a child, Lac Perras in Quebec. The lake was happened upon by a land surveyor for a logging company sometime in the late nineteenth or early twentieth century. When it opened for business, it catered mainly to wealthy Montrealers who would travel by train to Mont Laurier, (there was no practical access by road at that time) before embarking first by truck to the village of Lac du Cerf and then over the remaining several kilometers by packhorse to Lac Perras. It was usual in those days to stay at a fishing lodge for perhaps a month or even longer, and fishing was considered only part of a larger package that mainly involved escape from an (conceptualized as unhealthy) urban environment. Today comparable ritualistic displays of elite social status might involve owning vacation properties in exclusive locations around the globe. The perceived benefit in the early- to mid-twentieth-century Montreal was not entirely fantasy. By North American and even global standards, Montreal at the time had very elevated rates of infant mortality. The surveyor, Mr. Wester, as the story goes, found no fish in the lake during his initial investigation and stocked the lake by bringing in fingerling brook trout on horseback. He returned a few years later to find that the lake offered excellent fishing for large brook trout and decided to start a business as a *pourvoirie* or fishing and hunting outfitter.

When I started going to Lac Perras in the late 1960s, the lake had become much more accessible and was attracting an increasingly international clientele. My father had started fishing Lac Perras in the late 1950s or early 1960s with my uncle. They would always book the last week of May and the first week of June to coincide with an important hatch of large caddisflies. The exigencies of the school calendar precluded my brothers and I from those fishing trips, (although stories I later became acquainted with about those trips made the inclusion of youngsters highly unlikely in any case). One important aspect of place is its accessibility, and this aspect can flavor the whole experience of a place. This was certainly the case for me when my father invited me to go fishing at Lac Perras.

We would travel by car from a town near Syracuse, New York, on a journey over highway from Lafayette, New York, to Ottawa, Ontario, and then over increasingly country road, (the last 60 or 70 kilometers on dirt road). The trip took about six hours from departure in Otisco, New York to arrival at the village of Lac du Cerf, and then by large military transport vehicle to travel the last 10 kilometers from the village to the lake. In the 1970s, the dirt road up to the village was paved, making Lac Perras even more accessible and altering its nature as a place to go fishing substantially. The travel required to visit Lac Perras was very much a part of the experience of place, whether by rail and packhorse or by car. We often stopped by the side of the road between Ottawa and Lac du Cerf for a roadside lunch, and the smell of evergreens, roadside wildflowers, bird sightings, and fresh crisp air would always add to the feeling of anticipation. My father suffered greatly every spring and summer from hay fever starting in mid-June and lasting until the killing frosts of autumn, but there was one village along the road between Ottawa and Lac du Cerf which, once passed, would cause my father's hay fever to disappear entirely, literally within a few minutes as we traveled through the village. The hay fever would reoccur on the way home as soon as we hit the southern end of the same village. Often, trucks overloaded with logs and barreling down the dirt roads in the opposite direction would provide jolts of adrenaline. Even in the last days of June, when we arrived, the road to the camp would usually be wet, and the winch on the front of the truck would be needed to slowly inch our way through the forest.

With the forming of a majority government by the Parti Quebecois in 1976, large American fishing-club territories began to be nationalized. The pourvoirie of Mr. Wester and his sons was not entirely nationalized, but was leased to the Westers on a long-term basis within a much larger territory that did become a Provincial park. (The story the Westers told was that they wrote to the Minister and threatened to burn the forest if their business was nationalized and that story, coincidentally, illicited enthusiastic sympathies from many wealthy American clients.) Increasingly, over the years, the Wester brothers were required to adhere to Provincial standards with respect to the running of their pourvoirie business. These changing norms of behavior had implications for the attitudes of the guides and the attitudes of their clients, shifting again the role this place played in the constitution of identities as consumers of elite fishing experiences. At that time, there were two other outfitters also operating on Lac Perras. The conventional wisdom that emerged among the pouvoiries on the lake with respect to management of the fishery could be characterized by the term *put and take*. One of the Wester brothers once told me that the key was to continually introduce genetic diversity to avoid inbreeding, but there is no scientific basis for such a claim. My uncle held a very high position in the fisheries section of the New York State Department of Environmental Conservation and had often argued against stocking fish in Lac Perras. The stocking introduced disease from the hatchery into the lake and as years went on, the average size and health of the fish caught declined markedly. I have not returned to Lac Perras for well over 25 years now. Shortly before my last visit, rainbow trout were introduced, a move my uncle argued would only result in the further deterioration of the fishery. The Westers eventually sold their share in the business, and shortly after that walleyed pike were introduced by the remaining pourvoirie operators. That species would be expected to force the trout into deeper water, out of range of the fly caster. Recreational fishers generally do not have much control over the quality of the places they fish, unless they form associations with the goal of lobbying governments for such protection. Even then, they must be careful to always frame their lobbying arguments in economic terms and often any competing interest with higher economic potential will *ceteris paribus* result in the deterioration of the fishery. Those with low levels of political and economic power have generally seen a marked decline in sport-fishing opportunities over many decades in North America.

The long, steady decline of the Lac Perras fishery is representative of a broader trend across North America. Even in fisheries with more enlightened policies, where no stocking is allowed and severe restrictions are enforced on the numbers of fish killed, the quality of the fishing has declined. There are many contributing factors that explain this trend. On some rivers, for example, fishing pressure is so severe that wading fly-fishers literally crush substantial food sources such as caddis or mayfly larva simply by wading in the stream. Broader changes such as the introduction of alien species such as earthworms, chestnut blight, Dutch Elm disease, emerald ash borers, Eurasian buckthorn, Asian carp, or lamprey eels, as well as previously unexperienced toxins, or climatic patterns can take a long time to cascade through an ecosystem. Changes occurring over several decades may be difficult for humans to detect, especially if species variety appears to be mostly stable. However, it is difficult to say whether, for example, the disappearance of a once-dominant species such as American Chestnut, or the appearance of an aggressive species such as didymos, gypsy moth, or Asian carp might currently be having very broad repercussions for the entire biome of eastern North America that could still be cascading through various levels of ecological communities (Corlett, 2016; Shen et al., 2016). Again, it is not very informative to conceptualize the ecology of a place as being composed of many individual species. A wolf sharing a territory with musk ox is very different from a wolf sharing a territory with rabbits and deer. A given species of tree in an urban environment is not the same as a specimen in a complex forest community, and a polar bear on James Bay is not the same as a polar bear in a German zoo.

Often the experience of fishing places also responds to seasonal or temporal changes. An invitation to go ice fishing is quite different from an invitation to fish the same place in summer. Ice fishing is often a more social experience, since conversations involving other ice fishers, dog walkers, cross-country skiers, or other outdoor enthusiasts are much more likely to occur on ice than over open water. Ice fishing for walleyed pike is often done at night by groups of fishers in portable huts with camp stoves where chili and hot dogs are cooked. In the morning, they may display their catches to daytime fishers like trophies, eager to engage in conversation about the night's events. Many insect hatches occur only at specific times of year. Spawning cycles of both fish and the baitfish can also contribute to seasonal variability. Weather and climactic patterns also contribute to an experience of place with a heightened awareness of such variabilities. I have witnessed a few times when the fishing was unexplainably outstanding for several minutes just prior to a severe storm.

Narratives

Stories or narratives often act as the glue that holds specific triadic systems (parent-child- place) together. Our minds are particularly well adapted to putting the world into the form of stories (Clark, 2002, pp. 1–38; King, 2003), and this is especially true of fish stories. Some stories we inherit, and some, for example, evoke a dutiful sense that taking children fishing is an integral part of the implicit socially constructed identity of parenthood. Others we encounter, and some require no words: a child exploring the neighborhood creek creeps to the edge of an undercut bank and peers into the water to discover the orange, black, white, and green of a brook trout fin undulating in the current underneath.

Narratives do not always require words, and words enlisted to tell them may remain completely inadequate to their full expression. Like tools, narratives work to constitute identities. In any capitalist society, activities and events are generally sorted into categories of productive work or consumptive leisure. Learning to fish might in some very special circumstances be considered an investment in labor productivity. A salesperson who can strike up a conversation based on a fishing picture or stuffed fish on a wall will be more efficient at his or her job than one who cannot. As a rule, though, non-commerical fishing is considered a nonproductive or leisure activity. In this way, activities are cognitively woven into pre-existing cultural narratives. Other societies and cultures may sort these activities, events, or ways of relating into different conceptual categories, whether religious, mythological, or scientific, or by using some other seemingly arbitrary criteria.

This constitutive aspect of narrative results in a kind of pedigree of human experience. The world is experienced only through various mediators of materiality but is held together in place with a narrative of one sort or another, connecting past with present and future and providing the foundational understanding of the way the world works that is requisite for human action. Stories reflexively rely on specific configurations of materiality for support. Materiality and narrative are mutually co-requisite and partially co-constitutive. Fishing stories are about events involving material things, including fish, fishing gear, water, weather, and so on. A favorite fishing pole or boat, for example, can be a touchstone to anchor the meaning of fishing stories for the participants.

When a parent invites a child to participate in an activity, the stories are generally pre-existing for the parent, but only imagined by the child. The child may not have ever gone fishing before, nor even encountered fish as living things to be caught. My son, John, decided to share his snack of cheerios and raisins with the fish in the boat on one our first fishing trips together. The story emerging for the child from the invitation will probably have more to do with the possibility of acquiring a skill deemed worthwhile by the parent. Such imagined stories lean heavily on the innate psychological-developmental process that is constitutive of the human family. It is in our human nature to learn and to desperately want to learn such stories. Many of the cultural narratives we learn in this way are consciously articulated, but at the same time, their cultural meaning remains well below conscious awareness. One such narrative is foundational to Western civilization: the imperative to demonstrate control over one's environment, including fish. What follows is a fish story to illustrate this point.

One day about 35 years ago, I was fishing Oatka Creek in New York State when I came upon a fellow fisher standing on a gravel bar next to a large pool. The fisher was casting a mayfly imitation across the current to a pod of perhaps fifteen to twenty brown trout (two or three appeared to be quite large). A quick survey of the situation made me realize that it might be worthwhile to simply wait for this fisher to become frustrated and leave so I could take a turn at these actively feeding fish. I sat down under a tree at the bottom of the pool and waited quietly. There was a good hatch of mayflies bobbing along on the current from the rapids above and being swept by the current into the pool. Most mayflies would take flight from the pool, some actively fluttering their wings before lifting off. Some would drift through the pool to begin bobbing along in the next set of riffles below the pool. The rise form of the feeding trout did not match the activity of these mayflies. They were neither gulping large flies floating still on the water, nor splashily chasing fluttering mayflies about to escape. The rise form was more characteristic of a dimpling rise form that one would expect with a hatch of chironomids. This was in fact the case. There were two hatches occurring simultaneously. The other fisher was making two mistakes. First, the fly being used looked like a size 12 or 14 dark Hendrickson, which would have been a good choice to imitate the mayflies. The mayflies and the mayfly imitation were however being completely ignored by the trout. The second problem was with the way the fly was being presented. Casting across the current is a good technique when trout are taking skittering caddisflies, and although there were a few mayflies fluttering briefly on the surface, the majority were simply drifting quietly. To cast across the current to a specific trout under such conditions would require the shaking of some slack into the line during the cast, but this was not being done. The fly would begin to be dragged across the surface as soon as it landed. I fastened a size 18 gray midge on the end of my leader and waited. As predicted, the other fisher soon grew frustrated and waded to the opposite shore and began removing expensive chest waders. I immediately slipped into the bottom of the pool and cast the midge directly upstream and watched as it began a drag-free drift toward the location of a very large trout. Seemingly out of nowhere, a spunky brown trout of about 30 centimeters (12 inches) rose to the fly, struck, was hooked, jumped high above the pool, dislodged the tiny fly with a splash, and disappeared into the darkness of the pool. Instantly, the entire pool went silent, and it was clear that it would now be pointless to continue fishing that pool for at least an hour or two.

I am not exactly sure why that specific fishing experience has remained so close to the surface of my memory for so long. After all, not a single fish was even landed. I have no doubt that the story would have been quickly forgotten, had it occurred within a different cultural configuration. If the culture-based focus had been, for example, on fish for eating, then the experience would have been registered in the column under "failure." If the focus had been on social learning, it also would have counted as a failure. Why did I not suggest to the other fisher to try a chironomid imitation, or to shake some slack into his line for a better presentation? Where did the competitive aspect of this event come from? If the other fisher had not been there, would the experience have been so memorable? Did the event involve a ritualistic display of mastery on my part? What role did the place (Oatka Creek—a well-known and prestigious fly-fishing destination) play in the resonance of that memory? What role did the dominant cultural narrative of consumerism and the constitution of consumerist identity play? Perhaps the most pertinent question for this discussion is: What was it about the previous parent-child-place triads of my own psychological development that acquired me as that type of fisher? These questions pertain to the constitutive aspects of the normal psychological developmental processes; they in turn work to constitute specific features of real physical geographies, and these material configurations in turn have concrete implications for the sustainability of the entire assemblage.

Scientific knowledges have in recent history also worked to create the building blocks for these narrative types of experiences and events. Science produces evidence-based, or propositional, forms of knowledge, but just as we experience any event in narrative form, these scientific knowledges are ultimately strung together more or less adeptly and, for better or worse, come to be experienced as stories. Many fisheries, both sport and commercial, have been decimated in recent history at the hands of overconfident biologists and unchecked and misunderstood overfishing by factory trawlers. The Atlantic cod off the Canadian east coast is one famous example, but the Great Lakes also now represent only a faint shadow of their former productivity as a fishery. The introduction of Pacific salmon has decimated the forage fish of the Great Lakes, leaving little food for other sport-fishing species. These assemblages of unsustainability cannot be reduced to the human action of fishing, but more accurately reflect the cumulative impact of many human activities, including agriculture, forestry, industry, tourism, government, fish hatcheries, fish stories, and the various elements of many other assemblages as well.

New stories or narratives emerge from experience. My experience fishing Lac Perras as a child was quite different from my experience fishing Oatka Creek or the spring-fed pond near my home. Lac Perras worked to frame fishing as something done in far-away locations on special occasions. It involved skill sets that would be tested in unexpected ways, and often luck seemed to be the most salient factor in determining how successful the fishing would be. Would the weather be good? Would the flies hatch? Would the fish and their current daily routine make themselves known? The fishing I practiced daily near my home in New York State was quite different, and the story of fish and fishing that emerged from that experience was also quite different. I maintained a connection with the fish. I knew their changing locations and their habits on any given day at the place I fished daily throughout the year.

Process

Processes implicated in any invitation to "go fishing" can be as disparate as geochemical rules governing the spread of pesticides in hydrologic systems, or the ways increasing levels of mercury in fish are framed within a given cultural-political context. Processes are better understood from an interdisciplinary perspective as parts of whole systems, rather than from the reductionism of specific individual academic disciplines (see, for example, Berkes, Colding, & Folke, 2003; Berkes & Folke, 1998).

The invitation from a parent to a child to go fishing can form parts of several distinct processes. The psychological development process of the child is informed by the parent or caregiver's behavior. The invitation becomes a clue for the child regarding what is important in the world. If the fishing destination is operated as a commercial venture, and many children accompany their parents there on fishing excursions, then there will be more pressure on the outfitters to stock more fish, unleashing a chain reaction cascading through other processes. If enough parents take their children fishing, there may be some political pressure to stop or at least mitigate the negative impacts of logging, or industrial activities, near valued fishing places. These types of processes are subtle and often escape public awareness and concern.

If we read historical accounts of the stock of fish found by Europeans arriving in North America in the seventeenth century, we can only conclude that they have not been managed sustainably. The bounty that once was is now almost unimaginable. Yet identifying the processes causing such dramatic declines is not always easy. Certainly, the hydrological and ecological consequences of the dramatic deforestation negatively impact fish stocks. Yet deforestation is itself a symptom of a more extensive change in the regime of material engagement. Cronon (1983) describes this change as follows:

> Quite simply, the colonists' economic relations of production were ecologically self-destructive. They assumed the limitless availability of more land to exploit, and in the long run that was impossible. Peter Kalm described the process whereby colonial farmers used new land until it was exhausted, then turned it to pasture and cut down another tract of forest. "This kind of agriculture will do for a time," he wrote, "but it will afterwards have bad consequences, as everyone may clearly see." Not only colonial agriculture, but lumbering and fur trade as well, were able to ignore the problem of continuous expansion. When the gift was finally exhausted, ecosystems and economies alike were forced into new relationships: expansion could not continue indefinitely. (p. 169)

But the process of expansion continues. Mountains have been moved to produce "clean coal." Entire worlds are now populated with previously unimagined anthropogenic nanoparticles. Plans are underway to export people to Mars. These enactments of mythological narratives are continuations of the same story that previously transformed the Mediterranean from temperate forest to dry savannah (Harrison, 1992). Few processes can be contained within individual minds, but rather tend to jump across domains, from place to body or from brain to place. It is important, when seeking out the traces of (un-) sustainability, to consider processes from a cross-scalar perspective.

Agency

At first glance, it seems that the parent exercises the agency by inviting a child to go fishing. But if a grandparent had given the child a fishing pole, does the grandparent or even the fishing pole exercise some agency? If the parent's partner needs some quiet time to grade papers, does agency lie with the partner or the school board that employs the partner? If the outfitter phones to say the mayflies have begun to hatch, should we consider the mayflies the primary agent? Might the fishing place, or perhaps the fish themselves, be regarded as agents of the invitation?

Agency is an important aspect of sociocultural analysis and MET because it exposes the self-deception embedded in dominant accounts of the primary agency assigned to rational actors and arguments. This dominant approach does not provide an accurate map of the world but rather leads us astray.

Things happen all the time. There is, in fact, never a time in the human world when nothing is happening. Sometimes actions or events penetrate our conscious awareness and sometimes they do not. Either way, the things that happen around us become part of our experience of being in the world, even becoming the context within which our genes express themselves in the proteins that form our flesh and bones, to say nothing of the neural connections in our brains. It is the configuration of the landscapes, the inherited cultural/cognitive tools, and the routine communities of practice we engage in that ultimately exercise agency. Would I have ever gone fishing if my father had not invited me? Perhaps not, but the agency also originates in the physical and social environment that cooperatively steers developmental processes into a practice. Who can say what brought me to believe that writing about the paradox of a growing unsustainability amid increasing human knowledge was more worthwhile than planting trees or going fishing?

Memory

Memory can be traced, however tentatively, in brains, bodies, artifacts, institutionalized or ritualistic behaviors, or the natural or anthropogenic physical environment. Memories may become evident and available for examination in economic institutions, monuments, soil profiles, DNA, or many other psychological, cultural, physical, and/or material repositories of past events. At any given moment, the context for sociocultural action or material engagement includes the existing memories distributed throughout the brains, bodies, and landscapes (i.e., extended, distributed, and situated *minds*) involved. Memories fade and emerge continuously. Cognitively speaking, we remake the world continuously, including memories and all the other aspects of our minds. Even dreams can sometimes permanently shift our experience of the world (see, for example, Kohn, 2013; Lear, 2006). Fishing is an ancient activity, and in some parts of the world, more than mere traces remain to connect present practitioners with ancestral identities. The same could be said for foraging for mushrooms or picking berries in uncultivated places. For the most part, though, these memories are fading. Food gathered or harvested directly from the land and waterscapes comprises a decreasingly significant part of the average Canadian's diet. Climate change and prevailing patterns of land-use change will ensure that this trend continues, as traditionally valued and harvested species fall by the wayside (Minakshi, 2015).

Producing knowledge of the past for the sake of increasing knowledge is a quite different enterprise from remembering good knowledge for the sake of remaining human. Memory is important, in part, because it helps to establish cultural expectations and social feelings of rightness. Many Indigenous peoples of North America would consider catching fish with a hook and line to be disrespectful to the fish. The fish, if correctly asked, might be willing after all to sacrifice his or her body to feed the people. The right way to invite the fish might be to build a ladder of sorts for the fish and to help them navigate some rapids on their way to spawning redds upstream. The ladder could be constructed in such a way as to make the harvest of migrating fish easy, but the model in the mind is one of invitation for the fish to enter not into a trap but rather into a reciprocal relation with the people. This type of knowledge is accumulated slowly over many generations, but Western science is more interested in scientific breakthroughs with high market value.

Atmosphere

> We may think of atmospheres as the personal experience of a situation that may be qualified as "friendly," "cosy," "dull," "tense," or "depressing" and various degrees of "thickness" and intensities. Yet an atmosphere is never exclusively a psychological phenomenon, as state-of-mind, nor solely an objective thing "out there," as an environment or milieu; atmospheres are always located in-between experiences and environments ….
>
> Atmospheres are indeterminate above all as regards their ontological status. We cannot be sure whether we should attribute them to the objects or environments from which they proceed or to the subjects who experience them. (Bille, Bjerregaard, and Sørensen, 2015, p. 32)

The invitation from a parent to a child to go fishing will generally imbue the place where the invitation occurs with a specific atmosphere, albeit probably not with the same "thickness" as many other important proposals, such a proposal of courtship or marriage. The act of parents fishing with children also combines elements of bodies, tools, narratives, place, agencies, processes, and memories to also produce distinctive atmospheres. Memories rely on the smells, colors, and sounds of particularly resonant moments. Often the *look* of the water—the light, temperature, turbidity, and the wind—will remind me of a fish that appeared when the water looked the same.

There is a very sparse hatch of small black caddisfly that occurs at midday in the early spring on the pond I fished as a boy. I had been fishing the pond for several years before I made the connection between the occasional splashy rise and the infrequent skittering caddis in early spring. In fact, because the hatch occurred at lunch time and was so sparse, I am not sure how many times I would have had the chance to witness it before I finally learned how effective blind casting a cricket fly and fishing it as a caddis could be during these usually bright sunny noontimes in early May. Once aware of the hatch, though, it became an annual rite of spring. The atmosphere of spring fishing involves the attention, the seeking, of a splashy rise, or a small black skittering caddis at midday.

Fishing camps also have peculiar atmospheres, and not one just determined by the personalities of the people present. Sometimes the weather or whether the fishing is good also exercise only minor influence over the camp atmosphere. There are rules and protocols. For example, if the priest always needs to pray when it is time to wash the dishes and sweep the floors, then he may become the target of some good-natured joking. People who catch no fish after days of fishing but still stay in good spirits are generally much admired, especially if they also listen patiently as everyone else gives them advice on how to remedy the situation. The hope of a big fish generally continues to paint the atmosphere until the last rod is safely packed away in the car for the ride home.

Sustainability

Any serious discussion of sustainability should ask what it means to be human from within the specific culture in question. Our relationship to the (incorrectly conceptualized as separate) "environment" is determined largely by how we define and understand ourselves because our image of self mirrors our image of environment. What are the characteristics of the relations between the self and the not-self? How is that relationship conceptualized? Is the boundary between self and not-self porous or nonporous, permanent or ephemeral, universal or selective? These kinds of questions can help identify the cognitive/cultural tools in use as they pertain to the challenge of sustainability. Other questions can be designed to better understand different elements of the whole assemblage of (un-) sustainability. As we put these pieces together, other questions arise. What is it that makes the issue of sustainability crop up periodically as a critical issue for some specific cultures at specific times and now for the whole of humanity? How is the issue of sustainability framed when it does emerge? Does the issue resonate broadly or is it systematically suppressed, as, for example, it was in the case of the Island of Nauru (McDaniel & Gowdy, 2000)? The tiny island nation was inhabited happily and peacefully for roughly three thousand years before potash was discovered and mining combined with the introduction of firearms and alcohol destroyed the life of the original population.

Many other species of fauna (and some flora) catch and eat fish. What is it about the way some humans go fishing that is fundamentally different from the way otters, eels, or kingfishers go fishing? The main difference, I think, and the first key to understanding the challenge of sustainability, is the unprecedented amount of cognitive/psychological development that is postnatal in humans. This includes, but goes well beyond, the acquisition of language. Much of human development occurs in the same way it does in many other mammalian species—that is, through imitation and through simply being in the world. However, the system of representation and the cultural tools used to interpret those representations, whereby something comes to represent something else, sets the human species apart and must be negotiated with the very greatest of care. What, for example, do the words *fish* and *tree* and *market* come to represent within the context of English-speaking Canadians?

Clearly, there is no direct link between fishing and (un-) sustainability. Fishing is always nested within an array of other life activities that sometimes interact in complex and surprising ways. Indigenous peoples in North America harvested fish sustainably for millennia before the European invasion, but it would be pointless to examine Indigenous methods of fishing in isolation from a broader way of life. Their method often involved fishing one place intensively using gill nets and then letting that place rest for an extended amount of time that would allow its full recovery. The logic of reciprocity informed the practice and was also applied to hunting waterfowl. If these birds needed long runways for taking off from a lake into the wind, then removing some trees in their flight path could strike a bargain between hunter and prey, whereby the birds get an improved flight path and the people get meat. This method would seem to make infinitely more sense than the hybrid capitalist-style "environmental management" model of Lac Perras. But this shows how it is important to remember that systems need to be treated as systems and not mix-and-match collections of separable components. In the currently dominant neoliberal system, profit maximization is assigned the status of a trump card and this status is ingrained in the synaptic structures of the social mind, that web of frames, metaphors, narratives, vocabularies, material artefactes, and so on that structure social cognition and discourse. The most ecologically sustainable component might be inserted and quickly overwhelmed by the memory of neoliberalism (Berkes & Ross, 2016, p. 187). The marketing frame is like a highly aggressive invasive species, spreading broadly until the entire universe seems to be filled with only commodities and yet-to-be-commodified resources. I do not claim that at Lac Perras overstocking fish was the sole or even the most important cause of the decline of this fishery, only that it was a contributing factor and that these various elements of the entire context need also to be considered to have a good understanding of why this fishery was not a sustainable source of large brook trout. The clear-cutting for pulp wood around the lake certainly did not help, and as years went by, these types of changes to the land enabled hay fever to follow my father, and later myself, all the way to Lac Perras. In the case of North America, unsustainability arrived in the brains, bodies, and material artifacts of the invaders, but quickly spread to the entire system. It may be less than useful to consider such individual components in isolation.

Another key to sustainability lies in the reciprocal relationship between organisms and the environments. There is *never* just a human. There is no authentic self. The parameters of who we can become are set by the places we inhabit and that inhabit us, including all the material and cognitive tools therein. "There is no 'blueprint' for behaviour in the brain, or in the genes—no more than there is a blueprint for flock in the head of the bird" (Clark, 1997, p. 40). Furthermore, we cannot deliberately *design* ourselves by designing the places we inhabit and the places that ultimately constitute us. The mere act of assuming control makes the process fall apart. If the frames and metaphors put into social-cognitive practice are based on a dominance/submission model rather than a shared identity model in which the person I am becoming includes, and is integral to, the landscape that I co-constitute and that co-constitutes me, then sustainability simply will not get off the ground. There is, in other words, no "blueprint" for sustainability. Sustainability is, in one sense, an innate and, sooner or later, inevitably terminal problem for *homo faber*. The toolmaker will almost inevitably arrive at a state of mismatch between ecological processes and available tools that lead down an evolutionary cul-de-sac. Whether the origins of this mismatch of unsustainability are located within the changing aspects of physical places or in the culturally shared ideas used to operate in those places matters about as much as whether the chicken or the egg came first. Once a tipping point is passed, it can be extremely difficult to bring sustainability back, as evidenced here by the declining fish stocks of North America.

It is interesting that the gill-net fishing of the Cree "fits" with sustainability more than the fly-fishing practiced by wealthy settlers. One might think that fly-fishing brings the participant more into the world of the fish than gill-net fishing would. Fly-fishing requires a knowledge of the fish and of the fish's habits and food preferences. Gill-net fishing, on the other hand, requires only a knowledge of the fish's movements and the subsurface structures that might cause the fish to pass a certain location. However, once again, the point is how specific practices are interconnected with wider modes of thought and meaning. As Kohn, (2013) notes:

> Lives and thoughts are not distinct kinds of things. How thoughts grow by association with other thoughts is not categorically different from how selves relate to one another. Selves are signs. Lives are thoughts. Semiosis is alive. And the world is thereby animate. (p. 99)

Cultural/cognitive and material tools get woven into the social mind (see also Rappaport, 1968). The fly-fisher engages in an instrumental process of catching fish and, although part of the goal of that process is to constitute the person as a fly-fisher, the relational aspects of the process recede from consciousness precisely because of the specific tools acquired through the normal psychological developmental process. The Indigenous gill-net fisher has been acquired by a very different ontological framework, namely one that prioritizes narrative forms of knowledge over propositional forms and one that focuses more on relations than on objects. (Un-) sustainability is the product of a historical inheritance put into action through normal social and developmental psychological processes.

References

Benedito, S. (2017). *Landscape tunings: An urban park at the Danube*. Cambridge, MA: Harvard University Press.

Berkes, F., Colding, J., & Folke, C. (2003). *Navigating social-ecological systems: Building resilience for complexity and change*. Cambridge, UK: Cambridge University Press.

Berkes, F., & Folke, C. (1998). *Linking social and ecological systems: Management practices and social mechanisms for building resilience*. Cambridge, UK: Cambridge University Press.

Berkes, F., & Ross, H. (2016). Panarchy and community resilience: Sustainability science and policy implications. *Environmental Science & Policy*, *61*, 185–193.

Bille, M., Bjerregaard, P., & Sørensen, T. F. (2015). Staging atmospheres: Materiality, culture, and the texture of the in-between. *Emotion, Space, and Society*, *15*, 31–38.

Clark, A. (1997). *Being there: Putting brain, body, and world together again*. Cambridge, MA: MIT Press.

Clark, M. C. (2002). *In search of human nature*. New York, NY: Routledge.

Corlett, R. T. (2016). Plant diversity in a changing world: Status, trends, and conservation needs. *Plant Diversity*, *38*, 10–16.

Cronon, W. (1983). *Changes in the land: Indians, colonists and the ecology of New England*. New York, NY: Hill & Wang.

Ferraro, E., & Reid, L. (2013). On sustainability and materiality, *homo faber*, a new approach. *Ecological Economics*, 96, 125–131.

Harrison, R. P. (1992). *Forests: The shadow of civilization*. Chicago, IL: University of Chicago Press.

King, T. (2003). *The truth about stories: A native narrative*. Toronto, ON: Anansi.

Kohn, E. (2013). *How forests think: Toward an anthropology beyond the human*. Berkeley: University of California Press.

Kuhn, T. S. (1996). *The structure of scientific revolutions* (3rd ed.). Chicago, IL: University of Chicago Press. (Original work published 1962).

Lear, J. (2006). *Radical hope: Ethics in the face of cultural devastation*. Cambridge, MA: Harvard University Press.

Malafouris, L. (2004). The cognitive basis of material engagement: Where brain, body and culture conflate. In E. DeMarrais, C. Gosden, & C. Renfrew (Eds.), *Rethinking materiality: The engagement of mind with the material world* (pp. 53–62). Cambridge, UK: McDonald Institute for Archaeological Research.

McDaniel, C. N., & Gowdy, J. M. (2000). *Paradise for sale: A parable of nature*. Berkley: University of California Press.

Minakshi. (2015). Climate change, biodiversity and environmental challenges. *International Journal of Recent Research Aspects*, *2*(3), 122–124.

Rappaport, R. A. (1968). *Pigs for the ancestors: Ritual in the ecology of a New Guinea people*. New Haven, CT: Yale University Press.

Schmidgen, H. (2014). *Of actants, forces, and things*. New York, NY: Fordham University Press.

Shen, X., Bourg, N. A., McShea, W. J., & Turner, B. L. (2016). Long-term effects of white-tailed deer exclusion on the invasion of exotic plants: A case study in a mid-Atlantic temperate forest. *Plos ONE*, *11*(3), 1–16.

Vygotsky, L. V. (1978). *Mind in society: The development of higher psychological processes*. Cambridge, MA: Harvard University Press.

Wertsch, J. V. (1998). *Mind as action*. New York, NY: Oxford University Press.

· 4 ·

ECONOMIES OF MIND

> Bourgeois political economy—child of eighteenth-century thought—articulated a view of nature as "resource" and attributed to itself the prime theoretical task of determining the rational allocation of resources that were scarce. To this end it appealed to the theory of markets, to the goals of maximizing utility, and to the centrality of money as the common means to measure heterogeneities of human desires, of use values and of elements and processes "in nature." (Harvey, 1996, p. 150)

A fishing excursion, a walk in the woods, shopping at Bloomingdale's on 59th Street, or shopping at Bloomingdale's via the Internet in your pajamas, picking though trash bins for bottles and tins to exchange for money, planting milkweed seeds in the back alleyway while thinking of butterflies—these are all examples of activities that cannot be accomplished without culturally transmitted, mastered and mastering, mediational means. Ever since hominids began banging two rocks together to make tools, and through that practice becoming person-and-tool assemblages, the possibility of an unmediated experience of the world has effectively been foreclosed. The next two chapters focus on the origins, development, and consequences of the two dominant organizing characteristics of the Western-Cartesian historical inheritance, namely, an implicit cognitive and emotional focus based on economism and anxiety. Economism and anxiety form the basis of a tool set that transforms us into a cultural organism unsuited for sustainability and potentially unfit for longer term survival.

The banging of two rocks together was not a singular event. In fact, our ancestor, *Homo erectus*, banged rocks together to roughly the same end for well over a million years. At some point, though, that very act of routinized banging provided the evolutionary context for new kinds of banging, and weaving, digging, making, throwing, building, situated learning, bodily expression, thinking, feeling, talking, and otherwise becoming. Although the origins, processes, and structures of (un-) sustainability may be broadly knowable, and to some extent even chartable or representable through conceptual modeling, there is no final cause of (un-) sustainability. Rather, there are many situated, dynamic, mediated, extended, and distributed parts that may become thoroughly saturated with (un-) sustainability and are inter-related in many diverse ways.

I am not arguing here for Lorenz's chaos theory, whereby butterflies' wings might cause (un-) sustainability, but rather asserting that things tend to operate, and need to be understood, as whole systems. Human systems include agencies operating in fluid environments providing dynamic affordances and constraints, and freedoms and foreclosures. The moment-to-moment contingency (as opposed to the machine-like certainty) of the way the world works should be obvious, but our Western-Cartesian heritage means we often fail to recognize this.

Awareness of (un-) sustainability (like ethical awareness or behavioral codes) is always a learned, mediated, and situated awareness. Students of micro- and macro-economic theory continue to learn, to use the same myths, metaphors, and econometric models to search hopelessly for the final cause or determining mechanism of sustainability. (In this they are rather like politicians who learn to promise peace through military spending,) The usual suspects include calculating maximum sustainable yields (MSY), capitalizing "resources," or using a "triple bottom line" approach to economic decision-making. The aim is to reduce the world to a calculative experience. Such training may leave the student impervious to evidence that refutes the exigencies of mainstream economic theory even overriding the speed bumps lived experience places in the way of unfettered economic abstraction. Yet reality's speed bumps do not confine the "progress" of economism within the narrative, language, vocabulary, or material culture, and these limitations often seem meager in comparison to a human propensity for cooperative delusion. The evidence of such delusion is overwhelming (Rist, 2010).

Consider a trip to the supermarket or a walk down a suburban sidewalk. What provides the cognitive structure, the mediational means, needed to make these activities meaningful and purposeful? For one thing, the participants have already been assigned definitions as consumers of commoditized food products or viewers of real estate through their participations in sociocultural action. The globalized unsustainability mind necessitates everyone's constant participation in market-based activity, resulting in what Polanyi termed the "marketing mind" (Polanyi, 1977, p. 5). The marketing mind has no known boundaries. It is not confined to the checkout line at the supermarket or the dotted line of the real-estate deed. It colors all our current perceptions, feelings, and thoughts. Of course, this has a history, for example, in the fourteenth- to seventeenth-century practices in Britain of framing land as the private "property" of ruling elites, but it has since spread to everything—from land, to body organs, to education as training for the labor market, to ideas as intellectual property, and to genomes as marketable data. Everything in the world is now framed as marketable property.

Like a tick waiting for the smell of a warm-blooded host, unsustainability hides in economics textbooks, mowed lawns, polyester pants, cellular phones, and automobiles, waiting to locate and combine with the psyches of unsuspecting passersby. The marketing social mind is remarkable in the ways it disables us from perceiving the limitations imlicit in the marketing mind's perception of the world. Unsustainability and its agents are camouflaged partly by a culturally induced blind spot and a pre-existing ubiquity that normalizes the pathogenic into a conventional wisdom. The tools acquired (or foreclosed), whether in school (Graham & Arshad-Ayaz, 2016) or simply acquired as part of normal acculturation, provide little access to the contemporary state of (un-) sustainability. Without some form of guidance, sustainability remains simply unthinkable from within the contemporary gestalt of economism. Yet neoliberalism constrains and opposes any demonstration of actual practices that might be situated, distributed, mediated, extended, and enacted into material culture, physical environment, and embodied routines of people. And theoretical (utopian) visions and rational argumentation alone cannot constitute adequate guidance for the social mind. It is true that, in the sixteenth and seventeenth centuries, many individual European invaders to Turtle Island were able, with guidance, to Indigenize their way of life and adapt in previously unimaginable ways, to an unfamiliar (even utterly incomprehensible) economic landscape. Many *coureur de bois*, for example, intermarried with Indigenous peoples and participated actively in such Indigenous systems of meaning. More generally, though, European invaders negotiated and established far less benign correspondences to overcome cognitive and emotional dissonance. Familiarity and comprehensibility were aggressively and forcefully imposed upon the land and its ecology (Round, 1999). This process involved making the landscape fit with imported aspects of the European social mind: The land, for example, began to be conceptualized as a remnant of the original broken (post-flood) world, without the "benefit" of (civilized) human improvement.

At the societal scale, any Indigenization was simply unthinkable within a settler social mind and its world-transforming projects. We might say that the land of Turtle Island was quite literally driven insane as it was made to correspond with a psychopathic set of mediational means introduced by a people whose practices transformed forest to desert, induced disease and pestilence, and routinely engaged in the most callous and brutal treatment of Others. The introduction, and centuries-long development, of the insanity of economism transformed physical landscape and material culture together with individual and social minds. The result has been ecological devastation.

Economism was woven into social mind and environment using the cultural tool "property," to order thought and behavior concerning the land and everything on it. Of course, the sacred texts of economics, which could be dated to the now obscure writings of the French physiocrats in the late seventeenth to early eighteenth century had yet to be written when this process became routinized. The iconic work of Adam Smith, *An Inquiry into the Nature and Causes of the Wealth of Nations*, would only come some decades later, in 1776. But the artifacts, tools, and materiality of economism and unsustainability were already in widespread use. An exported European materiality—in the form of seeds, tools, building designs, clothing fashions, and many other examples of Cartesian-based material culture—was forced upon a Turtle Island ecology from the time of invasion, beginning in earnest in the sixteenth century. The physical embodiment and materialization, together with the ritualistic enactment of key narratives, formed the dual pincers of change.

The initial invading Europeans found no earthworms in the soils they sought to utilize. Earthworms had neither survived nor been reintroduced following the glaciation of the previous Ice Age (Langor et al., 2014; Rogers & Collins, 2017). Without earthworms to increase the drainage of the soil, the entire ecology of the north-east of Turtle Island had evolved with a more humid soil regime; and the different food, shelter, economic, recreational, and meaning systems of the Indigenous peoples reflected this different ecological reality. The Europeans not only introduced (knowingly and unknowingly) earthworms into the lands of Turtle Island, but they undertook massive and extremely labor-intensive projects to drain the land of what they perceived as excess moisture. The introduction of worms, along with aggressive drainage measures, did result in a more Europeanized landscape (Eliot, 1748; Flavell, 1709).

The cognitive tools, smuggled into Turtle Island in body/brains and material artifacts, modeled the human-land relationship on the purported relationship between God and humans. God, the land, and humans were understood through the metaphor of the accounting system. A quantity of sins that exceeded the quantity of good deeds led to eternal damnation from God and/or starvation from the land, while the reverse led to paradise and good eating. (For representative examples, see Eliot, 1748; Flavell, 1709.) Following expulsion from the Garden of Eden, many Europeans understood the world as being in some sense broken (Merchant, 2004). The human bargain with God involved a duty on the part of peoples to domesticate and fix the broken world in exchange for re-admittance to the Garden in the afterlife. But if the "broken" land was to be made productive, then something had to be put back. Traditionally, Europeans put animal manure back on agricultural fields to maintain productivity; however, due to a general lack of farm animals on Turtle Island, fish—the most abundant and easily accessible organic matter—was widely substituted. The early North American economic pattern followed something like this: Find the most thickly forested and most easily accessible tract of land possible. Burn it to the ground. Collect the ashes and ship those back to England for sale as fertilizer. Use the proceeds from the sale of the ashes to gain title to the land. "Improve" the land by installing drainage ditches and adding fish. Plant the land with European species of food and medicinal crops. Empty the rivers of fish to fertilize the now rapidly eroding agricultural soil. When the land becomes exhausted, move on and repeat the process (Cronon, 1983; Round, 1999).

The "extermination of contagion" gestalt previously discussed, combined with the economistic-accounting framework, formed an ongoing pattern of unsustainability even as it developed into the marketing mind. Adam Smith's *Wealth of Nations* might even be read as formalizing the idea that the workings of the "invisible hand" (markets) were God's way of eradicating contagion. We might, of course, be tempted to interpret the contemporary neoliberal social mind itself through the metaphor of contagion, in which case we should note that the negative feedbacks expected from a normal relation between disease and host species have so far failed to operate. No "balance" has been achieved; a disease that always kills its host will eventually find itself without a habitat. There are many known parasites that affect the behavior of their hosts. Economism often behaves more like a parasite that changes the host in some way that is beneficial for the parasite's propagation. The host—society, in the case of economism—becomes a sort of zombie, persisting, but in a very much diminished state. What may be most interesting about the comparison, though, is the process whereby "we become what we pretend" (Bateson & Bateson, 1987, pp. 167–182). With the contagion metaphor being such a dominant cultural/cognitive tool, it should not be that surprising that the adherents of economism begin to behave like a contagion.

Soil, again, provides a good example of this: "Soil is not just an abiotic environment for plants. It is teeming with life—billions of minute and not so minute animals, bacteria, and fungi. The interaction between the biotic and the abiotic makes the soil a living system" (Smith & Smith, 2000, p. 110). Yet industrial agriculture establishes a Cartesian relationship with the soil, whereby soils become abiotic environments that are poorly suited to meet the needs of plants. Plants, after all, have evolved over millions of years within living soil communities, not in sterile test-tube environments. The economistic logic of industrial production, exemplified by Taylorism and by Ford's production lines (Doray, 1988), uses mediational means that transform things into objects and co-constitutive relations into market-constituting commodities. These types of relations become fatal to many soil organisms as monocultures increasingly consist of plants that begin to resemble orphans from Ceausescu's Romania.

Despite its obvious tensions, the economism of the neoliberal mind now seems almost unquestionable in any foundational way. When markets fall apart, as they always do, criticism is allowed only to attack the constituents and products of the market but not the markets themselves. Such market failures are re-envisaged as cycles of self-correcting "creative destruction" (Schumpeter, 1934/2005, pp. 57–94), a model that provides a theoretical failsafe mechanism for economism. It reassures participants that, no matter how obviously insane or destructive the operating system becomes, progress is inevitably just around the corner.

The financial crisis of 2008 challenged conventional understandings of the way markets work to self-regulate and provide optimal societal outcomes. The high priests of economic wisdom who had been charged with overseeing the stock and housing markets were hauled before a Congressional committee in Washington, DC and forced to admit that their most deeply held assumptions about the way the world worked had proven false. Federal Reserve Chairman Allen Greenspan, for example, admitted, "I made a mistake in presuming that the self-interests of organizations, specifically banks and others, were such that they were best capable of protecting their own shareholders and their equity in the firms" (Knowlton & Grynbaum, 2008). Self-interest, the requisite model of subjectivity for any capitalist economy, had failed to deliver the goods. Indeed, unbridled self-interest had caused the economic system to go off the rails and required massive interventions by the US Federal Reserve and by other governments to avoid complete collapse. As we have seen with Hardin's (2001, pp. 60–62) tragedy of the commons, narrow conceptions of self-interested and individual preferences form the very foundation of the capitalist social mind, supporting otherwise nonsensical theories of value, welfare, and property. These, in turn, depend upon Cartesian and delusional conceptualizations of the individual in a bubble. Yet, the capitalist economic system depends entirely upon unwavering societal-scale trust in the magic of self-interest, narrowly conceived, to deliver maximum value and maximum welfare.

With the brief exception of the Occupy Movement (McGowan, 2016, pp. 82–89), public understanding and economic discourse following the financial crisis of 2008 never strayed beyond the zone of proximal development of this unsustainable social mind. No new cultural/cognitive tools arose to fill the void made apparent by the crisis. The social mind simply absorbed the event without learning anything from it, and Trump's election to the office of President of the United States ensured an administration that would work actively to undo any incidental learning that might have taken place. The crisis was simply interpreted as an anomaly, the exception that proves the rule. Far from being exceptional, these types of failures and their dismissal are ubiquitous. As Blyth (2002), for example, argues, the lessons from the Great Depression have also been successfully "disembedded" from the American economic system and social mind, especially since about 1970. Such contradictory evidence is over-ridden in accordance with a neoliberal ideology, including corresponding cultural toolset and material manifestations.

The parallels to the challenges represented by a societal transformation to sustainability are striking. For example, Jaworska's (2018) analysis of discourses of climate change in corporate social responsibility (CSR) reports from various oil companies found that

> tackling climate change becomes then, from the point of view of the oil industry, an act of balancing energy supplies with environmental consequences ... no radical changes are proposed and this together with the ... focus on climate change as a global issue to be addressed by all stakeholders suggests that companies intend to continue, to a large extent, with business as usual. (p. 209)

The oil industry frames itself as one part of an omnipresent economic system, therefore absolving itself of any responsibility to make changes. By contrast, those who advocate for making changes to the economic system are often labeled as fear mongering. For example, Spicer and Bowen (2017) argue that

> in invoking human fears for the survival of the human species or, at least, human civilization, whether in the short term or the long term, many advocates of sustainability may be seen here as engaging in what some political scholars have termed a "politics of fear." (p. 305)

Several authors included in Geiselberger's (2017) edited collection of essays, *The Great Regression*, confirm that humanity is currently locked into a downward spiral instigated by neoliberal economism. The fall of the Berlin Wall in 1989 was wrongly interpreted as marking the inevitability of market capitalism, leading to the hyperimplementation of the dual pincers of globalization and neoliberalism. These two policy options are not compatible, at least in the overdeveloped world, as well-paying jobs for the bulk of the population are shifted overseas just as the state removes any vestiges of a social safety net. Now stir global climate change into the mix. Crop failures and extreme weather events will ensure a growing migration of people across the planet for the foreseeable future. These circumstances favor the election of politicians who scapegoat immigrants and promise a return to past "greatness." These circumstances explain, in large part, the electoral victory of US President Donald Trump, Ontario Premier Doug Ford, Jason Kenney in Alberta, and so on. All these politicians seek to benefit from a laissez-faire approach to climate change, and why would they not? The image of migrants at the gates threatening to take jobs away from people whose livelihoods are already precarious ensures the continuing fortunes of neofascist politicians.

The contemporary unsustainability of the current psychotic economic system emerged as a novel species of "bad idea" in the ecology of mind. This type of cognitive/cultural tool, like the "capitalist" mind, or "marketing" mind, presents us with what Bateson (1972/2000 p. 207) terms a "double-bind situation" which involves:

"1. Two or more 'persons.'" In the case of the American social mind, we have Liberals and Conservatives or Democrats and Republicans.

"2. Repeated experience." This is a pattern woven into the fabric of being and becoming which, for those within the Western epistemological and ontological universe, becomes an exercise in the enactment of the capitalist dogma in every waking moment.

"3. A primary negative injunction." In this case, the dogma maintains that nanny state economic policies will undermine self-interest as the primary driver of economic performance and interfere with the exceptional nature of America.

"4. A secondary injunction, conflicting with the first at a more abstract level, and like the first enforced by punishment of signals which threaten survival" (p. 207). If we consider the dominant icons of American conservative ideology (NASA, unequaled military spending, and the regular exercise of force, the family, heterosexuality, objections to abortion, and so on), then none of these examples are even possible to imagine, initiate, or maintain through the unrestricted workings of market forces. Indeed free markets themselves can't be imagined without someone to enforce the rules. Markets do not simply organize themselves as could be seen after the collapse of the former Soviet Union. There was no capitalist or entrepreneurial culture in those countries, and capitalism did not spring forth like dandelions in a pesticide-free lawn. Some organization needs to be enlisted to cultivate a capitalist culture. Nevertheless, these conservative mythic icons have achieved the status of being beyond questioning despite their existing in continual and stark contradiction with the discourse of neoliberalism.

"5. A tertiary negative injunction prohibiting the victim from escaping from the field" (p. 207). The aggressive use of rhetorical devices across all forms of media has been largely successful in framing anyone who would question the benevolent omniscience of the market, especially in the face of evidence to the contrary, as somehow un-American. The greatest collective effort currently being engaged in across North America is the project of fitting the contradictory events of the day into the neoliberal narrative. The success of this project is truly a wonder to behold and a testament to the extraordinary powers of the human imagination.

6. "Finally, the complete set of ingredients is no longer necessary when the victim has learned to perceive his universe in double bind patterns.

Almost any part of a double bind sequence may then be sufficient to precipitate panic or rage. The pattern of conflicting injunctions may even be taken over by hallucinatory voices" (p. 207). Millions of Americans consistently misperceive the evidence on a range of issues including health care, austerity budgets, or gun regulations because contradictory evidence simply cannot be processed by the (anti-) social mind. The self-interest double bind has also become seemingly impossible to escape from, especially by the 1980s with the instantiation and enforcement of a widely accepted there-is-no-alternative (TINA) mantra.

The contemporary economistic social mind provides many examples of the Batesonian double bind. The main point here is that capitalism is very productively considered as a problem of a hallucinatory social mind in that reality is hidden within the inherent distortions of the socially shared and enforced mediational means. The current globally dominant economic system needs to be considered as a psychosis of the social mind. (Bateson's [1972/2000, pp. 159–342] own interest in schizophrenia can be traced, at least in part, to this realization.) The capitalist subject is forced into a double bind and a paradoxical unfreedom, deprived of access to the individuality so cherished within the mythos of capitalism. As McGowan (2016) argues, the entire state of capitalism depends upon the contextually *induced* desire of the individual:

> True freedom is freedom in the face of the Other's desire—or, more properly, freedom from the Other's desire. Freedom is an indifference toward the desire of the Other that the subject has when it finds itself fully immersed in its own satisfaction. The free subject ceases to concern itself with the question of the desire of the Other and pursues its own satisfaction regardless of its relationship to the Other. It neither tries to follow the desire of the Other nor deviate from this desire. But capitalism has a profound allergy to this type of freedom and does all it can to ensure a preoccupation with the desire of the Other. (p. 47)

McGowan's "Other" might be understood as including the material culture and environment that co-constitutes society. In other words, almost everything in contemporary Western society is dependent upon an inherited ritualistic devotion to economistic self-interest induced by an (anti-) social mind that is dialectically opposed to such mythically defined identities and behaviors. The agency of the rationalistic self-maximizer actually resides in society and in a particular material culture/environment. What we desire and demand emerges from our always-sociocultural relations and the mediational means we master and that master us. We are always actually constituted through such relations with others, both human and nonhuman, vibrant and inert, but we demean ourselves through our culturally routinized commodifying of others. Not only does the model of subjectivity required by capitalism put us individually and collectively into a Batesonian double bind, but it also forms the basis of inescapable unsustainability and psychic distress.

This is where mediational means, in a capitalist context, has real and important philosophical implications. The "who we are" becomes inextricably entangled with the "what we are" (Smith, 2011, p. 135). What we *are*, unavoidably, is a species of tool users, and, as such, we are at least partly defined, constituted, and mastered by those specific tools we engage, whether by choice, simple convenience of availability, or through a sort of osmosis by being in the same place at the same time. *What* we are becomes, then, co-constitutive of *who* we are. Peoples alive before stone hand axes are constitutively different than those who use stone axes or iPhones. Peoples employing and employed by a world of double-binding omniscient markets are not well suited for sustainability.

The contemporary subject is in part constituted from the entire capitalist context, including material artifacts and physical environments. As already mentioned, the material origins of the capitalist mind took the form of land enclosures (Polanyi, 1944/2001). These enclosures severed that historical interconnectivity in which the fates of peoples were very consciously and materially intertwined with the fates of the lands they worked. The regime of private property and the monetization and marketization of land was no less dehumanizing than the monetization and marketization of labor. As Polanyi (1944/2001) aptly notes, "The enclosures have appropriately been called a revolution of the rich against the poor" (p. 37). The enclosures removed the commons from the social mind, and grounded people-place identities were no longer available to structure thoughts, emotions, or behaviors. Commons were a materially present refutation of the implicit assumption of scarcity and the psycho-logic of narrow self-interest. Without the commons as a working component of the social mind, Hardin's (1968) tragedy of the commons became a reasonable way of perceiving, feeling, doing, and thinking unsustainability.

However, this tragedy did not suddenly appear holus bolus, but rather emerged over millennia from the less than skillful operation and reproduction of domus-agrios dichotomized social minds. In some senses, the origins of the tragedy of the economistic mind can be traced to the agricultural origins of Western civilization, as Gosden (1994) explains:

> From the late Palaeolithic period onwards, wild resources are used extensively; plants and animals were removed from the wild and brought within the cultural sphere. These moves increased people's dependencies on a small range of species and brought about genetic changes in the animals and plants, which in turn made them dependent on people for propagation and survival. All of these changes took place within a symbolic context centring on the house, through which both nature and society were domesticated. The house formed a centre of symbolism, the site of paintings, figurines, and pots, which also reflected a central concern with death. The dead were buried in the floor of houses, domesticating the natural process of death, which was seen to have a whole set of fears attached to it .... The disciplining of these fears also helped to hold social forces in check, through demonstrating the power of society to control nature. (p. 158)

This mechanism was not, as theorized by Diamond (1999, see above pp. 12-19), simply a matter of geographical accident or a case of uneven distribution of resources, whether genetic or material. It was rather the way that materiality was converted into conceptual tools and social minds. The origins of the Western mind grew partially from an idea, the socially shared, cognitively organizing cultural tool of a fear of contagion and death that was then materialized and ritualized into the practice of internment within the house. This materialization gave social thought and socio-psychological development a specific form and direction, a very peculiar "bulk of life" (Gosden, 1994, p. 125). Unfortunately, that direction did not then, and does not now, "fit" with ongoing ecological reality. By burying the dead within the cultivated or tamed and civilized space of the house, contagion and death were symbolically and cognitively conquered, reinforcing and propagating the conceptual split between the subjective imaginaries of culture and nature, to later be formalized in sacred text by Descartes.

Animal husbandry, as Diamond (1999) rightly asserts, was practiced differently in the Western mind than elsewhere. However, Diamond's thesis—that geographical variations in plant and animal resources provided historical advantages to Western peoples—ignores the human developmental and social psychological aspects of the evolutionary processes he theorizes. If the social mind is using a cultural tool to split the home from nature, then the way agriculture and animal husbandry are practiced will reflect that dichotomy. The cognitive split between what is tame and what is wild, or the structuring of space into home versus wild, dictated the pattern of physical interaction with environment that led to the experience of repeated collective psychological traumas. As Diamond admits, contagion became a regular and recurring event for Western peoples but he fails to recognize that strategies to immunize society against that contagion flowed from routinized practices that were ecologically incompetent. Diamond also fails to recognize the social-psychological-developmental consequences of such repeated traumas. Contagion continues to follow from the material configuration implicit in the Western social mind within the ecological context of planet Earth. As already mentioned, crop losses due to pests continue to increase with increasing pesticide use, and the rise in life expectancies initially afforded by antibiotics may soon be more than erased with the induced rise of various new "superbugs" that have been artificially selected for resilience and resistance to antibiotics (Shiva, 2016).

The issue here is not simply one of an increased concentration of people in shelters, which occurred throughout the world from the Neolithic onward, the Iroquois longhouse being one example. The problem arises when the shelter is conceptualized as distinctly separate from environment as in the tamed-untamed split, as material artifacts and biota are then segregated and amassed in much higher concentrations than they otherwise would be. This process might begin innocently enough, but as people continue to use these material configurations and practices to structure their thoughts, feelings, vocabularies, and conceptual models, the process of segregation accelerates. The growing of crops or raising of animals within certain areas designated as civilized might be prescribed with the establishment of specific social customs, for example, walls or palisades might be constructed. Then the trouble would begin:

> Pests had become a major problem by the time of the earliest recorded human history, and pest management occupied a considerable part of early society's attention. Biblical writings refer frequently to plagues of locusts; the inclusion of locusts in the Old Testament as kosher animals was thought to have served as both a pest control measure and a dietary substitute for the grain their swarms consumed. Remains of grain-eating beetles have been found in vases in King Tutankhamen's tomb dating back to 1350 B.C. (Winston, 1997, p. 3)

The keeping of large numbers of domesticated animals with restricted mobility, the planting of monocultures of plants, and the storing of foodstuffs all induced an increase in the frequency and severity of traumas that further reinforced the framing of nature as the enemy. That outbreak of pestilence then forms the backdrop for the social-psychological development of mind (brains, bodies, and places) operating now within the psychosis of Bateson's double bind—operating as recurring cycles of collectively experienced psychological trauma. Framing the covid-19 virus as the enemy in our new war has recently become the norm.

This positive feedback process may not have operated in a unidirectional fashion but efforts, whether deliberate or incidental, to bring about a more harmonious mode of existence have been overwhelmed within the Western, Cartesian way of becoming in the world. Cantor (2002), for example, posits a psychological connection between the Plague and Italian Renaissance that followed:

> Perhaps it had some impact on consciousness, but the trend was in the direction not of more soul searching but of a more earthly and secular experience. Perhaps the Black Death weakened faith in traditional medieval Catholic spirituality and set off a quest for a deeper naturalistic understanding of human psychology and behavior and the expression of a more personal sensibility ....
>
> Somewhere in the common unconscious there may have sounded the tocsin that the Black Death was the symbolic psychological decline of the old world of the European Middle Ages. It was time to get on with the new and modern, given added legitimacy by renewed classicism. But like all cultural history, this is just an ephemeral and unprovable observation. (pp. 210–211)

Tuan (1982) identifies several examples of shifting models of subjectivity developing from the psychological parameters of the Renaissance but also heading toward increasing ecological dysfunction. The shifting sets of mediational means characteristic of times that followed large-scale collective trauma would generally drive and be driven by the Western schism between home and the wild: in this case, the increasing number of separate rooms in dwellings, more elaborate tableware and dining etiquette, and respective cultural/cognitive tools reflecting an increasingly disconnected and individuated model of subjectivity. This then is the overarching trend throughout the history of Western civilization:

1. A collectively experienced trauma induces the psychologically defensive response of conjuring imaginaries of domesticity, safe from the vicissitudes of a now conceptualized as separate wilderness.
2. That new praxis of separation creates new and diverse material and ideational configurations such as more separate rooms in buildings, property boundaries on the land, and the Cartesian distinction between thought and matter itself. Soon much of human action is engaged in various boundary-policing rituals, such as the academic disciplinary separation of medicine from theater, ecology, cuisine, or music.
3. These cultural/cognitive severing tools invite further ecological consequences and further trauma in the possible forms of food insecurity,

pestilence, social stratification, armed conflict, organized (ritualistic) oppression of humans or nonhumans, the materialization of trauma in such forms as increasingly horrific devices for torture or execution, ecologically insensitive forms of architecture, language, religious custom, and so on.
4. The cycle repeats in an increasingly ugly positive feedback loop.

The traumas that get worked into a historical inheritance are not primarily physical traumas but rather psychological traumas. A shortage of food, in and of itself, may not represent a traumatic experience, but a shortage of food for some people while others have plenty will almost certainly be traumatic. For example, an outbreak of disease among your crops or animals, while your neighbors remain healthy, would qualify as a traumatic experience if the societal rules governing the collective response to that situation tend to blame the victim. Although the starting point for Diamond's analysis of a non-uniform distribution of resources may have been correct, it was the level of ecological skill that some societies demonstrated in avoiding trauma to the human social psyche over time that allowed them to avoid establishing a historical inheritance of ecological psychopathy. Western society's initial ineptness in the spatial organization of the environment favored the recurring emergence of traumatic experience and was allowed to accelerate into an insanity that spread throughout the mind, including institutions, environments, and all manner of cultural/cognitive tools.

It is important to stress again that what I am getting at here is not something occurring exclusively, or even primarily, at the level of rational, language-based thought. These cycles of trauma master cultures through art forms and architectures, artifact and materiality, as much as through metaphor, frame, or paradigm. That said, these traumas, of course, are also worked into the structure of languages. Structural differences in languages reflect and emerge from experiences of different environments but also reinforce and magnify those differences, including by recognizing, or failing to recognize, different forms of agency when attempting to address problems. For example, Bateson (1968), discussing differences between English, Arabic, and Tagalog, noted that they would "seem to have a great deal of importance for the development of ideas of responsibility or of free will as opposed to determinism." (p. 137). For example,

> Languages vary in the extent to which they allow constructions of the type, "The cookie jar got broken," and in the extent to which they imply a specified, active

> agent who could or should take action. Americans rail at [Tagalog speaking] Filipinos because they do not seem to live with the constant question, "What could/should *I* be doing about this?" Events move forward without specifically implicating a single cause or agent. (pp. 136–137)

Bowers (2012), too, recognizes how "the metaphorical nature of language not only carries forward the insights, misconceptions and silence of earlier thinkers, it is also essential to communicating about how relationships are to be understood and acted upon" (p. 305). He, like Bateson, bemoans the way Western languages tend to frame "the individual as an autonomous moral agent." Wengrow (2004) further extends this point about the linguistic inscription of collectively experienced trauma to the material effects of written text itself:

> All engagements with text impose forms of bodily comportment and action (or inaction) upon readers and writers. In highly literate societies, these are usually learned and adopted through unconscious imitation, as part of the repertoire of social skills acquired during childhood. In social contexts, where access to literacy or particularly potent kinds of text is more restricted, the bodily techniques of reading and writing may be more consciously elaborated and regulated as a sign of initiation. Furthermore, texts engage both readers and writers in reciprocal relations with absent others. These relationships, though not always based upon direct bodily interaction, nevertheless have a social as well as psychological dimension. Accordingly, it is both valid and feasible to treat writing as an aspect of the agent-like properties of objects and material structures, located within rather than beyond networks of social interaction. (p. 267)

Language and texts are inextricably caught up in the historical composition of a world incited by collective trauma arising from inept ecological action. Certainly, by the time the sacred texts of economism were being written, the sociocultural Rubicon had long been crossed.

Economic knowledge should be understood within this dynamic of increasing systematic unsustainability. Economic knowledge—in the forms of ideas, practices, and materializations—is the Trojan horse that gets into the minds of all children of Western societies and invites them to participate in this paradoxical dance of death and foreclosure. Yet education, so often touted as *the* solution to our ecological problems, generally addresses neither abstract narrative knowledges as designed artifacts that continue to design (Fry, 2012) nor reflexive materializations as *agents* (Gell, 1998). Not surprisingly, education fails to provide any realistic exit strategy from the accelerating spin cycle of unsustainability. The problem is not one of knowledge per se, but rather of models of subjectivity, shared sensibilities, and cognitive tools that rely on misleading metaphors and provide poor maps of territories that spread and constitute the working parts of an ecology of mind—that is, brains, bodies, and materializations and physical configurations of human tools and environments. The problem, as I have come to understand it, probably originated in the material affordances of the land that were unskillfully put into action, but this is a chicken-and-egg question. At this point, any exit strategy requires the complete reconfiguration of our ecology of mind (defined minimally as operating systems of brains, bodies, tools, and places) to allow the land to take the lead and bring us and our tools back to socio-psychological health.

Economism, then, is not simply a story. It extends far beyond that. Economism includes the casual way of filling a gas tank without ever calling to mind the wars and environmental destruction that routine action implicates. It gets into the workings of our body-brains and imparts a direction and purpose to the tools we use. No tree or rock can escape its influence as it resets the composition of the Earth's atmosphere and climate. The opportunity costs of economism, though, fall most heavily on peoples and their foreclosed potentialities.

Bodies

Today capitalism transforms bodies both conceptually and physically. The way we think about and experience our bodies is always through the lens of capitalism. The activities we engage our bodies in tend to revolve around commodification. Whether laboring, recreating, learning, eating, competing, or any of the other various activities that form bodies through (in)action, capitalism currently provides the context that makes those activities meaningful for the majority of the world's population. "Though love emerges earlier in human history than capitalism," explains McGowan (2016), "it appears as if capitalism invented love *ex nihilo* given how useful it has become for fostering the subject's investment in capitalist relations of production" (pp. 176–177).

Bodies become billboards advertising self-control and improvement, exemplifying and extolling the rationalization of bodies into production units maximized for efficiency and competitiveness. Of course, people residing in food deserts have difficulty producing movie-star bodies from Cheetos, Whoppers, Seven-Up, and Kraft Dinner, but if they fail, and they have the cash, they might still achieve happiness vicariously by purchasing and playing a strategic simulation video game such as The SIMS, by hiring a dating service, or through some other dopamine-stimulating paid-for experience (Rifkin, 2000). There may be something different, too, about the experience of burning calories with an expensive fitness-club membership, compared to the equivalent energy expended in walking or backyard gardening; but gardening, hiking, or other seemingly heretical activities are still actively pulled into the logic of capitalism with the right spade or running shoes. They too are understood within the narrative of value maximization, so that a garden's purpose becomes intertwined with notions of property resale value and exercise as a means of increasing one's value in the labor market. Even leisure becomes an act of consumption, an act of value maximization within the capitalist cognitive universe.

Tools

The conceptual and material tools of capitalism are, as argued above, extensive and all pervasive; indeed, they are often shared across ostensibly competing tool sets. This is an important point if our goal is the achievement of real sustainability. Although the economic system of a market economy emerged from "an ingrained habit of thought" (Polanyi, 1977, p. 1) and not the other way around, it is important to note that this ingrained habit also militates against the development of genuine alternatives. It might be argued that various forms of socialism failed, partly because no equivalent tools were available to them, as for example, in the emergence of socialism in czarist Russia and the subsequent top-down imposition of Stalinism. Such failures do not mean that cooperation is a myth; it always exists, before, during, and after any particular economic formation. But where the dominant social institutions are produced in and through Cartesian-organized environments and according to the available tools of competition and instrumental command and control, economic forms like capitalism are difficult to dislodge. The "marketing mind" pre-existed capitalism, and no sharing-tooled mind, no historically inherited mediational means associated with cooperation and reciprocity were available for the start of Soviet communism. Indeed, both capitalism and socialism emerged from essentially the same Enlightenment historical inheritance. The history of czarist Russia was no more filled with love and kindness than was pre-industrial Britain. Both societies shared a worldview that devalued traditional Indigenous knowledges and sought above all relative greatness and dominance through industrial progress. Both political forms seem bound to the same, or only marginally different, Cartesian tool set, organizing and giving structure to their respective social minds.

This point again illustrates the agency of the already-lived-in world and its capacity to guide the emergence and development of economic institutions. There are, in fact, many capitalisms and many socialisms, each reflective of the mediational means and the already-lived-in world where they operate. However, although the ideologies of capitalism and socialism are today available to anyone with access to the Internet, the critical question is about resonance (see Chapter 1, pp. 19–26). (Un-) sustainability tools must resonate in order to be brought into person-tool-environment triads, and this is especially true of tools used to envisage and develop a shared understanding of the way the economy works. We must at least consider the possibility that the failure of the Soviet economic model was its failure to resonate within a broader and deeper historical context.

Place

Capitalist places are disciplined according to the psycho-logic of the marketing system. They become property to be disciplined by owners and quantifiably optimizable according to the one-dimensional capitalist metric: profit. Just as bodies serve the labor market, places, including all their human and nonhuman residents, must also prostrate themselves to the metric of the real estate or commodity market. Whether fishing lodge, family farm, automobile factory, coal mine, or Madison Avenue skyscraper, the capitalist aesthetic is a numerical expression of the return on investment (ROI). Capitalism becomes the only calculus through which hopes and fears become expressible, and place, within the Capitalist mind, is transformed into a set of symbols within that web of ROI calculus. Capitalist places speak their capitalism just as Weiss (1995) argues that French formal gardens in the seventeenth century reflected their own social and environmental conditions.

> Every object, cultural or natural, is caught within a web of symbols—these symbols themselves are altered, sometimes ravaged, in the course of time. Furthermore, no symbol exists in a pure state: we bring to any symbolic system the "impurities" of our own understandings and misunderstandings, as well as the desires of our most utopian hopes and cataclysmic fears. (p. 9)

Just as the French formal garden symbolized a world under the complete control of the king, under capitalism, land more generally became a symbol for world under the absolute control of the invisible hand, as Polanyi (1944/2001) explains:

> It can be readily seen that market economy involves a society the institutions of which are subordinated to the requirements of the market mechanism.
>
> The proposition is as utopian in respect to land as in respect to labour. The economic function is but one of many vital functions of land. It invests man's life with stability; it is a site of his habitation; it is a condition of physical safety; it is the landscape and the seasons. We might as well imagine his being born without hands and feet as carrying on his life without land. And yet to separate land from man and to organize society in such a way as to satisfy the requirements of a real-estate market was a vital part of the utopian concept of market economy. (p. 187)

Narratives

The capitalist story is an all-encompassing general theory of everything, a cultural evolutionary adaptation to the millennia-long origins of recurring anxiety and trauma. It is also a story of magic in which markets can turn knowledge into wine, or fishes into loaves, with no more than a slight bit of theoretical friction. The stories of capitalism are double-binding stories. They assure us that no matter how destructive the process becomes, you could be the next "slum dog millionaire," and in any case, there is no alternative story. Such stories work their way into our identities and compose our lives, making it difficult for people to evaluate narratives that effectively structure their daily routine behaviors.

Consider the story of self-interest as explained by Hardin (2001) and note especially the agency of this subplot within the grand narrative of capitalism:

> A common complaint against rational choice theory is that it mistakes concern with self-interest for rationality. The complaint is well grounded but pointless. Rational choice theorists commonly do suppose that the agent's own welfare is the chief value of concern to the agent. But there is no mistake involved in such an assumption. It is rather a hypothesis than a direct finding. If it produces good predictions and explanations, it is a good hypothesis; otherwise it is a bad hypothesis. In any case, it cannot be judged a mistake merely by direct inspection of motivations. The critic should want to show not that the assumption is wrong by inspection but that it leads to wrong predictions and explanations. (p. 57)

Here we have a nice illustration of a failure to be reflexive concerning theory that greatly facilitates the double bind of economic stories. The problem here is not that the ridiculous economic assumptions about human nature do not "produce good predictions and explanations," but rather that they do produce a tautological template for those behaviors. The selfishness assumed within economic theory is not, in other words, given by nature, but rather internalized from the stories of economic theorists and the material configurations that play supporting roles. The linear models of the Cartesian worldview are partially successful only because they obscure the double-binding loops of (un-) sustainability.

Processes

The processes of capitalism include the mundane ways in which a dominant ideology/culture always exerts a strong influence over the perceptual framing practiced by its inhabitants. In many societies associated with various varieties of traditional knowledges, the world is alive: literally everything in it has life. In societies acquired by capitalist epistemologies, the world is for sale: everything in it can be known in monetary units of account. The process of learning to perceive market mechanisms as ontological agents is the necessary and sufficient prerequisite for the emergence of the capitalist economy. That initial developmental process in turn leads to others. The host of the parasitic marketing mind can disregard increasing deforestation or pollution, while still maintaining both a firm belief in the overall progress of science and technology and a hallucinatory perception of a continuous upward ratcheting of human welfare. The processes of human perception and cognition, under the veil of mainstream economic thought, have been rightly called "delusional" (Rist, 2010).

This process of becoming a "marketing mind" is the same process that *could* become a sustainability mind. Some vicious cycles can transform into virtuous cycles. The origins of the process of capitalism were primarily material. According to Polanyi (1944/2001),

> To antedate the policy of laissez-faire, as is often done, to the time when this catchword was first used in France in the middle of the eighteenth century would be entirely unhistorical; it can be safely said that not until two generations later was economic liberalism more than a spasmodic tendency. Only by the 1820s did it stand for the three classical tenets: that labour should find its price on the market; that the

> creation of money should be subject to automatic mechanism; that goods should be free to flow from country to country without hindrance or preference; in short, for a labour market, the gold standard, and free trade. (p. 141)

These institutionalized ideas (laissez-faire, commodity, market, etc.) were preceded, in other words, by the hard facts, the material proof, that land, labor, and money *were* commodities because those physical things were already active co-conspirators in the "great transformation" to a market-based economic system. Yet Polanyi also argued that the transformation could have gone the other way in Western Europe because of a long historical experience of commons. If the process of transformation is consistent, then the enlistment of things as co-constitutive of mind might also nudge the system back in the right direction.

Again, this transformative process might be best understood and enacted through a relational ontology, as opposed to the disjointed and mechanistic Cartesian ontology. A relational ontology might shepherd an economic system into the world based on a person-environment relationship where each extends into the other, rather than one in which each is reduced to an exchange value. A relational ontology might weave a world into existence in which one's own welfare is recognized in the welfare of the community, in the already-lived-in world where the parameters of who a person *can* become are set.

It is worth underscoring the fact that there is no wizard behind a curtain, pulling levers and turning dials to operate the economy; there is no "invisible hand." An economy is path dependent and emerges from the already-lived-in world as each generation "decides" which tools to carry on with, which ones to reject, and which new ones to create (Hodgson, 2001).

Agency

This point brings us to the hallucinatory nature of plastics, perhaps the material most iconic of the era of neoliberalism. Plastics provide an unparalleled example of the ways in which agency is distributed in double-binding contradiction to rhetorical consumer sovereignty. It is not a simple task to delineate the various actors and their capacity to extend influence. As Malafouris (2013, 221) reminds us, "Action involves the coalescence of human and nonhuman elements, and thus the responsibility for action must be shared among those elements. No distinctions between human and nonhuman entities can be sustained in terms of agency." Imagine, for example, a gentleman of a certain age who is losing his ability to hear or imagine a diabetic. The hearing aid or the hypodermic needle requires plastics for its manufacture. Deafness is not a life-threatening illness as diabetes can be, but in each case, provided one lives in a caring society where medical services are available to all people, it would be generally expected that these conditions be treated and treated by enlisting the use of plastic materials. There is, in the treatment of hearing loss or diabetes, plenty of agency to go around. The patient, the doctor, the prescription, the device manufacturer, the device, the pharmacy, the family of the patient, the (economic) expectations of the society, the affordances of the plastic material, and many other actors exercise agency within these processes of plasticizing environments.

Although the extent of plastic pollution today may be reaching a point that impinges on future health and livelihoods, it is impossible to locate the "cause" of the problem and apply remedial measures. People who make plastics do cause plastics to accumulate in environment, but are not entirely in control of that process. In the making of stone tools, "knapping is what causes the movement of the hand, the changing of the stone, and the neural activation pattern in the parietal cortex of the knapper. Knapping, then, is the true agent and controller" (Malafouris, 2013, p. 175). In the making of plastic medical devices, it is in this now largely economically structured process of "acting upon" where the agency originates. It is in the mastering of mediational means where the agency to enlist, make, design, manufacture, and distribute plastic medical devices that comes to seem as normal as eating chickens, taking showers, or breathing.

Memory

Memory is blurred by capitalism (Hodgson, 2001). The importance of historical context is diminished as the world itself is perceived and thought by the capitalist subject to have operated according to eternal and universal market principles since the time of the Garden of Eden. A kind of cost-benefit analysis is woven into many of the metaphors, myths, parables, and paradigms we use to make sense of the world. Applied to memory, something must first be considered worth remembering within the capitalist mind before effort can legitimately be assigned to the task of remembering. In the abstract world of economic theory, where resources can be combined according to the recipe du jour with knowledge and transport costs and then mixed and matched to suit any occasion, there is not much point in becoming too attached to any specific time, place, or resource setting. Markets move rapidly, and labor must be prepared to go wherever the market sends it. Sentimental souvenirs simply impede efforts to achieve greater efficiency. The connection between memory and identity is squeezed by capitalism. Rational economic actors need to take stock of the here and now, the current market conditions. Data from the past is useful only for extrapolating into the future, while decision-making is always based on current or present-valued prices, converting even time into a unit of account.

Farrarese (2017) reflects on the lived experience of capitalism as a form of socially organized forgetting:

> Full experience, or *Erfahrung*—experience anchored in the possibility of linking history to one's personal life—is replaced by atomized *Erlebnisse*, or brute isolated experiences, experiences that are more like convulsions or traumas than segments of intelligible existence. They become accidents from which one has to be defended. In particular, the organization of work in a factory alters each one of the senses, isolates the worker from all experience, replaces memory by conditioned responses and savoir faire by repetition. Practice no longer has any signification. Such an atrophy induces a profound mutation in one's relation to oneself, to time, and to nature, but above all to others. (p. 177)

In this way, capitalism renders us less vibrant as memory, activated and deactivated with the Western-Cartesian mediational means, develops a reset button as each moment in time involves a recalibration and re-attunement to the current market conditions. This applies as much to the demand side of the capitalist world as it does to the production side described by Farrarese. Every act of consumption resituates us in the marketed world, our own market value recalibrated as we jockey for competitive advantage. To borrow Adam Smith's famous example, it is the framing of the world as a mathematically solvable equation that drives the pin maker to convert the pin-making process from an artisan endeavor to an industrial one. Then, as the market conditions for pins changes, rational consumers and producers will simply get out their calculators and adjust their buying and selling practices accordingly. There is no use in crying over spilled milk, whether the result be the end of an industry or that the river by the pin factory is now full of toxic by-products. Anyone foolish enough to attempt to include unquantifiable sentiments into their revised calculations will simply be disciplined by the market. In other words, when the market functions as envisioned by its proponents, we transform ourselves from healthily connected person-tool-environment triads to one-dimensional rationalizers with self-inflicted memory loss and consequent loss of identity.

Atmospheres

What can we say about the atmospheres of capitalism? Some examples of capitalist mind-capitalist world associations would include the frenzied buying and selling of the stock market, the frenetic pace of life lived under a cloud of continuous competitive anxiety and of choking city smog, the consumerist fervor that pervades Black Friday, the noise and clamor of the factory floor, or just the constant clicks of office information technology, the sense of always needing to update that online presence, and the atmosphere of constant surveillance and manipulation. We also should not ignore the chemically altered air that we breathe, the crowds we inhabit in ironic isolation, the tangible, and, above all, the intangible and implicit sense of a world, supposedly omnisciently disciplined, yet obviously out of control.

This is not to say that the atmospheres of capitalism form an unfragmented and uninterrupted habitat from which there is never any escape. Quite the contrary, the paradoxes of capitalism continuously open fissures in the experience of capitalism. The double-bind aspect of the capitalism gestalt ensures that life is experienced as a series of frustrations and contradictions. But these contradictions are also themselves a constitutive aspect of capitalism's persistence. Success is supposed to follow learning to live and prosper within capitalism's oppressive and anxiety-filled atmospheres, to fit with and become fitted to its corrosive contradictory "nature." The psychic distress this causes is not necessarily so very different from that caused by concern that one's family might fall prey at any moment to the Inquisition, and the cultural/cognitive tools employed as coping mechanisms at that time are roughly the same as those that now enable the capitalist to actively participate in the ongoing ecocide. Today, instead of learning how to conceptualize our relatives as heretics or witches, we learn to conceptualize them as commodities whose destiny is subject to the dictates of the labor market instead of the Inquisitor. We also fail to conceptualize the entire family of life as our "relations" (see LaDuke, 1999).

Curtis (2013) accurately describes this current state of affairs as follows:

> Ordinarily, empiricism's demands for proof and verification challenges the authority of priests and mystics, but with the establishment a largely unaccountable corporate

> aristocracy who continue to propagate faith in the self-regulating character of markets despite all evidence to the contrary, it is perhaps appropriate that we should be called on to simply acknowledge the supremacy of the market rather than interrogate the assumptions underpinning its mystery. (p. 74)

The experience of capitalism has become, in other words, the experience of participating in a society whose primary self-(re-)productive strategy is the normalization and stabilization of its own collapse. The dominant atmosphere of capitalism is one imbued with the precarity of all existence.

The paradoxes (contradictions) of capitalism can thus be stated as follows: In trying to produce an atmosphere of present and future prosperity, we have instead produced an impoverished world; the feeling of chronic precarity most now feel is only matched in extent by which the denial of precarity is necessary to generate continued adherence to capitalism. The atmosphere of intolerance to alternatives is only matched by a world that is becoming increasingly intolerable for ourselves and to so many other species. We inhabit an atmosphere that is fearful and anxiety-ridden. The historical experience of being in a garden, a shared intentional multi-species community, is increasingly pushed by capitalistic mediational means toward the experience of being (trapped) in a shopping mall (Merchant, 2004, pp. 167–184).

Sustainability

Capitalism is antithetical to sustainability at least partly because the two require mutually exclusive tacit knowledges, mutually exclusive epistemologies and ontologies, and mutually exclusive cultural/cognitive tools. Capitalism-peoples and sustainability-peoples simply represent two mutually exclusive ways of becoming in the world. The shared model of subjectivity required for a capitalist economy to function would be inconsistent with sustainability even if its only negative impact was to devalue life. But capitalism not only enforces the cognitive and emotional template required for the commodification of living beings and other agents (land, labor, and money were identified by Polanyi (1944/2001) as "fictitious commodities") (p. 75), it also contains an entire system of incidental poison pills. Plastic packaging provides one such example. Like so much of contemporary material culture, plastic packaging suddenly appears indispensable in the double sense that we can no longer get along without it and that even when its negative consequences become widely known, it still never goes away; from landfill to oceanic gyre, it continues to poison.

The next chapter will appraise of the current state of the loop-structured unsustainability processes as evidenced in the emerging screen culture. Our surrender to the somewhere else of the screen parallels previous social-psychological surrender to the promise of paradise in the afterlife, needed as a coping strategy to face the radical conceptual separation of the civilized from the barbaric, and the categorization of the world and everything in it as either master or servant.

References

Bateson, G. ([1973] 2000). *Steps to an ecology of mind*. Chicago, IL: University of Chicago Press. (Original work published 1972).

Bateson, G., & Bateson, M. C. (1987). *Angels fear: Towards an epistemology of the sacred*. New York, NY: MacMillan.

Bateson, M. C. (1989). Language, languages and song. In R. W. Rieber (Ed.), *The individual, communication, and society: Essays in memory of Gregory Bateson* (pp. 129–146). Cambridge, UK: Cambridge University Press. (Original unpublished conference paper written 1968).

Blyth, M. (2002). *Great transformations: Economic ideas and institutional change in the twentieth century*. New York, NY: Cambridge University Press.

Bowers, C. A. (2012). Questioning the idea of the individual as an autonomous moral agent. *Journal of Moral Education*, *41*, 301–310.

Cantor, N. F. (2002). *In the wake of the plague: The Black death and the world it made*. New York, NY: Perrenial.

Cronon, W. (1983). *Changes in the land: Indians, colonists and the ecology of New England*. New York, NY: Hill & Wang.

Curtis, (2013). Thought bubble: Neoliberalism and the politics of knowledge. *New Formations*, *80/81*, 73–86.

Diamond, J. (1999). *Guns, germs, and steel: The fates of human societies*. New York, NY: Norton.

Doray, B. (1988). *From Taylorism to Fordism: A rational madness*. London, UK: Free Association.

Eliot, J. (1748). An essay upon field-husbandry in New England. New London, CT: T. Green.

Farrarese, E. (2017). Precarity of work, precarity of moral dispositions: Concern for others in the era of "emotional" capitalism. *Women's Studies Quarterly*, *45*(3/4), 178–192.

Flavell, J. (1709). *Husbandry spiritualized or the heavenly use of earthly things* (10th ed.). Boston, MA: John Allen.

Fry, T. (2012). *Becoming human by design*. London, UK: Berg.

Geiselberger, H. (Ed.). (2017). *The great regression*. Cambridge, UK and Malden, MA: Polity Press.

Gell, A. (1998). *Art and agency: An anthropological theory*. Oxford, UK and New York, NY: Clarendon Press.

Gosden, C. (1994). *Social being and time*. Oxford, UK: Blackwell.
Graham, P., & Arshad-Ayaz, A. (2016). Learned (un-) sustainability: Bandura's bobo doll revisited. *Journal of Education for Sustainable Development, 10*, 262–273.
Hardin, G. (1968). The tragedy of the commons. Science, 162, 1243–1248.
Hardin, R. (2001). The normative core of rational choice theory. In U. Maki (Ed.), *The economic worldview: Studies* in the ontology *of economics* (pp. 57–74). Cambridge, UK: Cambridge University Press.
Harvey, D. (1996). *Justice, nature & the geography of difference*. Cambridge, MA: Blackwell.
Hodgson, G. M. (2001). *How economics forgot history: The problem of historical specificity in social science*. London, UK: Routledge.
Jaworska, S. (2018). Change but no climate change: Discourses of climate change in corporate social responsibility reporting in the oil industry. *International Journal of Business Communication*, 55, 194–219.
Knowlton, B., & Grynbaum, M. M. (2008, October 24). Two lions face a reckoning: Greenspan makes rare admission of fallibility. *International Herald Tribune*, p. 1.
LaDuke, W. (1999). *All our relations: Native struggles for land and life*. Cambridge, MA: South End Press.
Langor, D. W., Cameron, E. K., MacQuarrie, C. J. K., McBeath, A., McClay, A., Peter, B., … Pohl, G. R. (2014). Non-native species in Canada's boreal zone: Diversity, impacts, and risk. *Environmental Reviews, 22*(4), 121–136.
Malafouris, L. (2013). *How things shape the mind: A theory of material engagement*. Cambridge, MA: MIT Press.
McGowan, T. (2016). *Capitalism and desire: The psychic cost of free markets*. New York, NY: Columbia University Press.
Merchant, C. (2004). *Reinventing Eden: The fate of nature in Western culture*. New York, NY: Routledge.
Polanyi, K. (1977). *The livelihood of man*. New York, NY: Academic Press.
Polanyi, K. (1944/2001). *The great transformation: The political and economic origins of our time*. Boston, MA: Beacon Press. (Original work published 1944).
Rifkin, J. (2000). *The age of access: The new culture of hypercapitalism where all of life is a paid-for experience*. New York, NY: Jeremy P. Tarcher/Putnam.
Rist, G. (2010). *The delusions of economics: The misguided certainties of a hazardous science*. London, UK: Zed Books.
Rogers, J. A., & Collins, C. D. (2017). Ecological predictors and consequences of non-native earthworms in Kennebec County, Maine. *Northeast Naturalist, 24*, 121–136.
Round, P. H. (1999). *By nature and by custom cursed: Transatlantic civil discourse and New England cultural production, 1620–1660*. Hanover, NH: University Press of New England.
Schumpeter, J. A. (2005). *The theory of economic development: An inquiry into profits, capital, credit, interest, and the business cycle* (11th ed.). New Brunswick, NJ: Transaction. (Original work published 1934).
Shiva, V. (2016). *Who really feeds the world? The failures of agribusiness and the promise of agroecology*. Berkeley, CA: North Atlantic Books.

Smith, M. (2011). *Against ecological sovereignty: Ethics, biopolitics, and saving the natural world.* Minneapolis: University of Minnesota Press.

Smith, R. L., & Smith, T. M. (2000). *Elements of ecology* (4th ed.). San Francisco, CA: Addison Wesley Longman.

Spicer, M. W., & Bowen, W. M. (2017). Are you scared yet? On the ethic of sustainability and the politics of fear in public administration. *Public Integrity, 19*, 301–315.

Tuan, Y. (1982). *Segmented worlds and self: Group life and individual consciousness.* Minneapolis: University of Minnesota Press.

Weiss, A. S. (1995). *Mirrors of infinity: The French formal garden and 17th century metaphysics.* New York, NY: Princeton Architectural Press.

Wengrow, D. (2004). Violence into order: Materiality and sacred power in ancient Iraq. In E. DeMarrais, C. Gosden, & C. Renfrew (Eds.), *Rethinking materiality: The engagement of mind with the material world* (pp. 261–272). Cambridge, UK: McDonald Institute Monographs.

Winston, M. L. (1997). *Nature wars: People vs. pests.* Cambridge, MA: Harvard University Press.

· 5 ·

TRACES OF SOCIOCULTURAL ANXIETY

This chapter will endeavor to explain an influence on, and by-product of, the marketing economy: sociocultural anxiety. To begin this discussion, we need to recognize that, as already argued, any economic system entails specific mediational means embedded in the specific models of subjectivity operating and being operated by that system. The contemporary system of global capitalism is especially pernicious in this regard. What keeps an economic system in place is always a healthy dose of tacit knowledge situating peoples in ways they cannot fully express in words. Any adequate description of the dangers of capitalism is therefore difficult, if not impossible, to express in language alone. This is how unsustainability continues to penetrate the core of our being ever more deeply. We might also have to recognize the importance of what Skoll terms a shift from logocentric to iconocentric forms in late capitalism.

> The rise of iconic representation as the dominant form of communication and consciousness marks an even greater epochal shift than the transition from a capitalist to a postcapitalist world system of political economy. The shift from the logocentric to the iconocentric began approximately in the mid-twentieth century. The advent of television signaled it, but should not be confused as its cause. Image began to replace text as the preeminently authoritative sign. The iconic shift strengthened the effectiveness of the fear culture, as characteristics of icons lend themselves to more emotional and less analytic styles of thought. (Skoll, 2010, p. 115)

The rise of iconic representation associated with the explosion of media did not return us to the cave paintings of Lascaux. In a sense, we never left Lascaux, but rather have always uncovered or constructed meaning from iconic elements of environment. What has changed, however, is the nature of the parenting that environment now provides. Our interaction with a world increasingly constituted with plastics, highly processed and microwaveable food options, screen devices, mowed lawns, and so on has become less analytical and more emotional.

Certainly, fear can be summoned quite dramatically within genres of horror films and literature, violent video games, or far more maliciously as terrorism, but fear and its more mundane and longer term expression, anxiety, are also an aspect of contemporary existence that needs to be considered as a type of dominant background noise within a much broader ecology of mind. Skoll's point is that the current capitalist system constitutes a particularly dangerous mediating means, transforming the world we perceive into mere iconic representations as commodities, self-interested profit maximizers, buyer satisfactions, externalities, and so on. It is somewhat ironic, then, that the hero of the story, the rational "economic agent," is rendered nearly superfluous by the system itself. What the economic agent desires is often more incidental to an iconocentric subjectivity than it is to anything connected in any way to what we would normally identify as "reason." This ability of the economic system to redistribute intention and agency does not, of itself, trigger any generalized society-wide anxiety, but the continuous operation and exercise of that ability does.

This point about the near irrelevance of intent does nothing to diminish or undermine the magnitude of anxiety's potential agency or capacity for influence within the entire mental system of (un-) sustainability. In this sense, a "shift from the logocentric to the iconocentric," as described by Skoll, is certainly a troubling prospect that would only reinforce the urgency of the present thesis. Part of my argument up to this point has been that the logocentric form of consciousness has become increasingly delusional within the Western-Cartesian unsustainability mind. A shift to mental processes increasingly reliant on iconic representations, within the contemporary material culture and built-environment context, would only make the remedial paradigm shift less susceptible to consciously planned instrumental actions.

I agree with Skoll that fear is a powerful agent within the Western-Cartesian mind. While icons of fear are certainly prevalent on television and in other media, fear is also distributed nearly ubiquitously throughout the

Western (un-) sustainability mind, showing up as much in military budgets, mowed lawns, or exotic pets whose genetic design responds to deep-seated insecurities and obsessions with control, as in more purely cognitive tools that give structure, form, and tone to more mundane thought and speech. According to Vygotsky (2012), "Schematically, we may imagine thought and speech as two intersecting circles. In their overlapping parts, thought and speech coincide to produce what is called verbal thought. Verbal thought, however, does not by any means include all forms of thought or all forms of speech" (p. 94). Nonverbal thought includes aesthetic sensibilities, and some forms of tacit and embodied knowledge, but these forms of mastered and mastering nonverbal knowledge do not exhaust the category of nonverbal thought, either.

Economism, as I argued in the previous chapter, has invaded much of our conscious and subconscious, verbal and nonverbal, thoughts and feelings and has become woven into the fabric of psychological development of the Western-Cartesian child. Economism situates and orients our perceptual, cognitive, and emotional practices. As Chapter 4 suggests, if we dig a little deeper into the Western subconscious, underneath economism, there is an acquired and acquiring, implicitly learned tone or hue of anxiety. Whether the transmission and reception of that anxiety occurs via television, in song, art, cultural narratives, or any other mediational means, it has been a dominant feature of the Western-Cartesian historical inheritance for a very long time. Anxiety is evident in the confrontational approach of Western medicine or criminal justice and in many other places including, as we shall see, the sterilized appearance of homes and gardens. Anxiety drives space travel and cryogenics. That historical, culture-based anxiety forms the foundation not only for the Cartesian worldview and the "marketing mind" but also for child-rearing regimes, educational philosophies, art and music styles, and so on.

Certainly, anxiety is what maintains the paradox of capitalism, and this anxiety emerges within the shift to an iconocentric model of subjectivity. The real trouble arises, though, when the cultural tools of economism—seemingly designed on the back of a cocktail napkin after one too many martinis—bang up against real brains and real bodies in a real world. Surrendering to the iconic simply causes the trouble to sink to a deeper level of consciousness. This implicit fear characterizes and escapes from the Western-Cartesian mediational means and disciplines our perceptions, understandings, and feelings of ourselves and our world. Anxiety now also arises from the increasingly likely prospect that the mediational means that have come to constitute us could be exposed as a fraud, delegitimized, and banished to the realm of outdated and discredited mythologies as we find ourselves increasingly gasping for unpolluted air and drowning in an ocean of externalities. Paradigm shifts are generally painful affairs and, consequently, vigorously resisted. That background anxiety—fear of change and fear of the collapse of the self through the collapse of Western mediational means—deeply inures us to the obviousness of a world being actively driven to collapse.

This same root anxiety that disciplines people to burn forests to the ground for profit is a fear born from the illusion arising from a Western-Cartesian ontology, and it forms a prerequisite for the original emergence of that same worldview. For example, when we make land a commodity and then expect every landowner in a community to maximize the value of their "asset," only to later discover that the community has lost its soul along with its green space and its connection to a broader community of life, then the social mind will need some pretty good repression techniques. We might come up with, for example, a theory of the tragedy of the commons or a theory of negative economic externalities. Neither of these techniques, however, addresses the problem of the underlying anxiety.

The origins of anxiety as a constituent of the Western worldview can most probably be traced back as far as the Neolithic period and the specific ways in which agriculture and animal domestication emerged as shared, ritualistic patterns of practice. The instrumental approach to technological innovation as an organized response to underlying anxiety differentiated Western civilizations from most other coherent cultural ways of becoming in the world. As we saw in Chapter 1, that difference has been historically evidenced, clearly but not uniquely, in different approaches to and outcomes of agricultural and livestock activities. The practice of agriculture within the Western-Cartesian social mind came to reflect a relationship that was based on counting, separating, categorizing, adding, subtracting, and (initially at least) remembering—i.e., keeping score. Life became oriented toward the number of cattle or the number of bushels of millet or the number of trees in an orchard, keeping the "good" plants and animals separate from the "bad" ones, and always keeping track of the number of transgressions versus good deeds or bountiful harvests versus famines, and so on (Eliot 1748; Flavell 1709).

From the Neolithic to the present, the Western way of becoming situated in the world has included an obsession with objectified and fetishized knowledge in response to a deeply situating subconscious model of subjectivity oriented around anxiety. That anxiety-based sensibility suggests, frames, and institutionalizes the questions we ask, the methods we enlist in the answering of those questions, and finally, the answers to any question that can be recognizable as answers. It is not an overriding or all-encompassing terror; rather it acts like a background setting—a collectively learned and socially practiced anxiety that simply expresses one's mediational means. It operates at a subconscious level and becomes most apparent in its materializations. Those materializations—such as plastics, mowed lawns, antibiotics, pesticides, genetically modified organisms, space shuttles, disciplined knowledges, and countless other mundane examples—reflect an obsessive-compulsive quest for control. What could drive such a culturally embedded psychic template if not a deep-seated historical insecurity and fear of the unknown and the unknowable?

The contemporary wicked problems associated with, for example, nuclear waste and nuclear weapons do not preclude the practices of uranium mining or nuclear-power generation. Fearful things do not overcome us or preclude active, informed, and voluntary engagement with those things, mainly because they are conceptualized as being directed against a separate and distinct Other. The by-products of the fearful things and activities we willingly engage with are conceptualized as being disposable and removable to the safety of some abstract place where all scary things can be buried or flushed away and forgotten. The primary impetus for nuclear power, in any case, was weapons-grade plutonium and not energy for more benign uses. As Rhodes (2012) explains, "Power reactors were delayed in the years immediately after the Second World War because everyone involved in the new atomic energy enterprise believed that high grade uranium was rare in the world, too rare to be diverted from weapons production" (p. 41). Anxiety operates as a dominant pattern erupting at every opportunity in such individual instances or events.

Consider, for example, this critical difference between normal child-rearing practices of Australian Indigenous communities as contrasted with modern Western-Cartesian child-rearing norms. First, the Australian Indigenous child exists both conceptually and physically "in community" rather than as isolated object, as Byers, Kulitja, Lowell, and Kruska (2012) explain:

> Child rearing is shared by all who have a responsibility to the child within the kinship network. This extends beyond the biological mother and father to include all who have a relationship to that child, as well as those who have specific responsibilities maintaining the social structure of the community. (p. 295)

Within the non-Western world-making process, "Children grow up with constant engagement and inter-action in the community life" (Byers et al., 2012,

p. 295). In contrast, the Western approach enforces the conceptual framing of children as a separate and distinct ontological category rather than as members of a co-constituting community of life. The Western approach focuses on mastering the child as an apprentice "master-of-self" to later become a "master-of-Others." The importance of this distinction cannot be overstated:

> In this [Aboriginal] community, babies are integrated into community life from birth, included in community activities rather than being isolated or separated because of their age or lack of maturity .... This is in contrast to the segregation of Western children from the work and social life of adults, which reduces the opportunity children have to learn through participation. (Byers et al., 2012, p. 296)

How many Western-Cartesian adults are engaged in employment activities that would be appropriate for children to even witness? What would our world look like if children were present during all our world-making activities, framed as community projects rather than as the aggregate economic product of individual self-maximizers? These differences have little to do with verbal thought but rather more often operate below conscious awareness through the mediational means and situated tacit knowledges and aesthetic sensibilities that are ultimately more determinant in cultural-ecological world-making processes.

The obsessive-compulsive drive to expand the social-ecological psychological sense of security associated with the conceptualized as isolated, tamed, and the civilized, enables a repression of anxiety that would perplex the less delusional. Francis Bacon's role in the advancement of knowledge, for example, is still being debated (Merchant, 2008), but the ontological shift from participant to disinterested and separate knower that Bacon characterized and proselytized, as well as the detrimental consequences for sustainability that shift represented, are impossible to deny (Cosgrove, 2001). No doubt, the shift in mediational means from tools that framed us as being with the world to being in a world that is "like a container" began well before Bacon, but even very early "experiments" in child development reveal that detachment from a world regarded, or framed, as purely objective and lacking potential for social and emotional involvement has appalling consequences, as Hundert (1997) makes clear:

> As early as the thirteenth century, Frederick II, King of Germany, devised an "experiment" to see what language children would speak if he ordered their foster mothers to feed, bathe, and clothe them but never speak to or around them ....[His aim being to determine the original language spoken in the Garden of Eden.] Unfortunately,

> virtually all of the children died, and he concluded instead that the loving words of a mother are needed for human survival. (p. 5)

This situation parallels extreme contemporary situations in which the mother is alcoholic or the father is manic-depressive and is repeatedly hospitalized; infants do indeed grow up in a world where "objects come and go capriciously."

In such cases, the children often grow up to have an understanding of "objects," and even of other people, "that is very different from the one usually employed, especially with respect to the permanence and constancy of those objects" (Hundert, 1997, p. 77). As adults, they have a detached understanding of the world as somewhere full of arbitrary objects that is, in some respects, very like Bacon's. This is a fearful, capricious world in need of constant and compulsive management and control.

The experiment designed by Frederick II was probably the first recorded demonstration that the psychological development of human children is not simply about being thrown into a world in any arbitrary way, like some people still believe baby bears-in-zoos can be. Just as a bear-in-zoo is fundamentally different from a bear-in-the-wild, the child who has developed connectivity only to the strictly controlled parameters of a highly technologized and hyper-controlled habitat is not the same as sustainability peoples would be (see Louv, 2005, 2011; Turkle, 2011, 2015). The world is not a disinterested neutral container, simply forming an abstract isotropic plane on which human agency may be exercised with impunity. Children have some minimum requirements regarding the type of meaning system, the type of material and emotional environment, that can be tolerated. Human development occurs along a continuum with cognitive and physical atrophy leading to death—as documented in Romanian orphanages and the Frederick II experiment—at one end of the spectrum and a healthy connectivity resulting in sustainability at the other. The key seems to be connectivity, but the establishment and development of connectivity to unsustainability is not fatal, at least initially. Even child soldiers are still woven into something connecting brains, bodies, and places (Özerdem & Podder, 2011). Romanian orphans and the subjects of Frederick II's experiments, on the other hand, were denied even that minimal possibility (Hundert, 1997). The absence of the connectivity required to construct a potential meaning system causes brains and bodies to atrophy and eventually die, yet the brain-body will work with what tools are readily available and accessible, regardless of how hallucinatory or psychotic those accessible tools may ultimately prove to be.

What is currently believed, (and scientifically known), about the human psyche (a large part of the *what* we are) provides us with some basic guidance with respect to establishing sociocultural-ecological parameters on the spectrum from sustainability to unsustainability (the *who* we become), and eventually to contemporary constellations that prove to be incompatible with life. The normal developmental process involves bringing the psychological functions of the individual or society into increasing correspondence with the immediately perceptible landscape. Children learn the language they are exposed to. They learn what to expect from caregivers, what to expect from objects they come across, and what reaction to expect from the different behaviors, including facial expressions they practice. They learn etiquette and social norms, what to pay attention to and what to ignore, and what to care for and what to mistreat. Throughout this process, the mind searches for or invents a story or meta-narrative to connect everything together (Clark, 2002; King, 2003).

Something that does not fit into the story will generally feel uncomfortable and trigger the mind to search for a way of either ignoring the anomaly or knitting it into the accepted narrative as the exception that reinforces the general rule. For example, within the Western-Cartesian paradigm, an indigenous species garden can reinforce the imperative for control with demonstrated knowledge of the Latin names of all the plants and a high level of control over the color scheme and location of the plantings. The relationship between garden and gardener remains essentially the same as the relationship with the mowed lawn: the rational planner imposes abstract materializations of "rational" thought onto the conceptualized-as-blank-slate matter, something like the God-World relationship as described in Genesis 1:1 of the Bible. The garden continues to be experienced through the mediational means of the gardener, acting as surrogate for God, (or Francis Bacon) imposing order onto chaos. The replacement of non-indigenous with indigenous species in the garden involves simply replacing one expert knowledge with another. In this case, despite good intentions, the indigenous species garden can even be accommodated to the story of capitalism as a form of creative destruction driven by the knowledgeable exercise of consumer sovereignty. The garden then just provides propositional knowledge to reinforce the abstract, environmental-economics model of a sustainability that never quite seems to achieve any real progress.

Taken in isolation, then, even such exceptions can reinforce and normalize the more nefarious pathologies of a Western-Cartesian social mind driven by fear-based behavioral patterns that, in turn, induce ecological consequences that continue to reiterate and magnify yet more fear, anxiety, and terror. As Hinchliffe, Bingham, Allen, and Carter (2017) explain,

> In Western states, at least, emerging infectious diseases have become emergencies in waiting. The threat of a widespread and acute malady affecting people, or indeed the plants and animals on which they rely … is a long-standing one. But it is the imagined and to some extent experienced interdependencies and vulnerabilities that people share with each other that seem to have raised the stakes in the last few decades. So much so that we are, for some at least, "teetering on the edge" of a major disease event or catastrophe. (pp. 3–4)

So, while the means to provide pseudo-security (and usually corporate profits) at the micro level multiply, at the macro level, our collective subconscious plots our suicidal demise. How is this possible?

Western-Cartesian behaviors invite catastrophe, and pretentions of an evidence-based approach often merely camouflage the ways that conventional wisdom assigns misleading meaning to such evidence in ways that predetermine that the world will be experienced as a fearful place. Gifford and Gifford (2016) make a similar point in relation to the psychological reaction to extreme weather events induced by anthropogenic global climate change:

> Along with PTSD and complicated grief, other mental health issues persist beyond the immediate disaster into seemingly calmer times. Many of these issues are more gradual and subtle in nature, making them more difficult to measure. For example, some residents of tornado-prone areas have developed pathological fears of extreme weather events, so severe that they are afraid of leaving their homes. (p. 293)

This is the *normal operating procedure* of the Cartesian delusional inheritance. Anxiety is subconsciously nurtured and doted upon while double binding humanity and the physical-ecological world into an arms race with nature. Indeed, Cartesian-induced unsustainability could not be maintained without ritualistic enactments and material embodiments of the Western myth system that require the disinfection of everything. These materializations become the bridge where mind is extended and distributed across environment and environment is extended into mind.

Although material artifacts of the human landscape always "shape the mind" (Malafouris, 2013), a shared narrative is also needed to produce stable-enough structures to be recognizable as a working culture. The narrative glue of the current global order is economism, but economism still emerges from, and relies for its continuation on, a deeper subconscious sensibility of fear. The objectivity espoused by economistic forms of knowing is not a neutral or conceptually superior way of knowing the world, but rather the dominant instance of a Bacon-like attempt at scientific "containment,"—an attempt to impose order on that fearful world it actually has a key role in producing:

> Thus, ontology [from the perspective of the mainstream economist] can always be stated in epistemological terms. In spite of the philosophical language, this contains a warning based on common sense which economists should heed: the structures, institutions and actors of the economy cannot be ignored and treated as if they had no existence except as predictable functions or operations in the scientific train of thought. (Pfouts, 2002, pp. 71–72).

And yet they *are ignored*, and not just by economists or despite a continuing deluge of evidence to the contrary.

This conventional wisdom represents a specific set of mediational means emerging from a shared history of material engagements. Western science itself is in this sense a reaction to and a consequence of specific events, which when turned on itself implodes, collapses, and is exposed as simply a zigzag reflexive mirage. It is not that the scientific method does not produce useful knowledge; rather it is that it disguises that knowledge to seem like something it is not, namely ontological security and immunity from a reflexive and entirely interconnected world. Disguised as an alchemic ticket to immortality and complete knowledge, Western scientism is often the magic charm that transforms the world into predictable functions or operations, tamed, mollified, civilized, and capable of making the expectation of future trauma disappear. Why do we refuse to let go of Western scientism and not simply begin the process of honoring it with appropriate funerary rituals, perhaps some appropriate wailing and lamenting, followed by a burial and then a moving on to more sustainable and healthy ways of wondering, asking, knowing, making, and just plain finding our way? The short answer, I think, must be a broad-based and deeply felt sociocultural anxiety. The Western-Cartesian worldview provides a false sense of security—for which there is neither any ontological nor any metaphysical basis—but a shared sense of security (if not invincibility) all the same. Western science is the teddy bear that helps us confront a world now too frightening to be faced with the failing security blanket of Western religions alone.

We are now, and have been for many generations, in an arms race with our own mediational means. Our attempts to take control of the world are creating an increasingly inappropriate context for human psychological development. This applies as much to the artifacts with which we surround ourselves as it does to the landscapes we discipline with capitalist logics and to the minds we discipline with expert parenting advice or formal schooling. The available models we provide ourselves and future generations with are increasingly impoverished and frankly provide misleading guidance for developing sustainable understandings of ourselves and our world. Just as there can be no winners in a nuclear-arms race, there can also be no winners in the Western-Cartesian quest to reform the world in the image being dictated by our mediational means.

With the benefit of hindsight, the knowledge produced (at such appalling cost to lives) by Frederick II represents a historic missed opportunity. Perhaps if only some cultural tools that were not fear-based had been available to interpret the findings of his language-in-the-Garden-of-Eden experiment, we might have followed a different, more sustainable, path. The experiment demonstrated conclusively that child and environment are co-extensive. If the importance of empathy and connectivity had been recognized, perhaps the flaws of Cartesian dualism, for example, could have been recognized, isolated, and resisted even as they developed. There are many examples of such lost opportunities, including, for example, the eradication of Celtic cultures that, with their place-based understandings of sacred springs and groves, herbal remedies, and so on, represented a kind of truce with nature that Cartesian-minded Christians had earlier taken exception to.

Giambattista Vico's *The New Science* (1968), published in 1744, similarly represented another missed opportunity (Le Moigne, 2013). Vico's methodological clarity and sanity clashed too directly with the dominant fear-of-contagion gestalt. Vico, as Harrison (1992) explains below, had an intuitive understanding of the concept of mediational means:

> Vico had argued that history is determined by gradual mutations in the customs, laws, institutions, beliefs, languages, and modes of thought of peoples as a whole. Vico spoke of an "ideal eternal history traversed by the history of all nations." His theory of "Poetic Wisdom" held that the modes of thought of ancient peoples were essentially poetic, not prosaic. This notion of Poetic Wisdom became essentially critical for a historian like Michelet, concerned as he was with questions of national character and popular wisdom. Michelet spoke of Vico as the "golden bough" allowing him to descend into the realm of "infernal shades" and resurrect the spirit of the past. And sure enough, the way back into the realm of the past led to the forests of remote antiquity—the forests where Vico searched for the origins of Poetic Wisdom. (pp. 165–166)

Sacrificing realism and self-understanding for an imaginary psychic security, Western society followed Descartes instead of Vico and continued down the rabbit hole into current unsustainability. While other cultures managed to survive without an inherited mediational means resulting in wars, pestilence, and ecological impoverishment, the Western cycle of fear and terror continued unabated—from the enclosures to the Industrial Revolution and to European modernity—without much awareness of any problem. A mere four decades after Vico's *New Science*, Malthus's *An Essay on the Principle of Population* (1789), ensured that the unparalleled misery of the common people in a period of agricultural and industrial revolution was not questioned and interpreted by Poetic Wisdom, but rather extended with the analytic tools Malthus and others provided. This cultural tool set disregards misery and suffering and sets the stage for the events of Auschwitz, Hiroshima, Bhopal, Abu Ghraib, and Love Canal—terrible events resulting in fear and fear resulting in a search for more control, which led to more terrible events. The sense of security provided by the Cartesian perspective simultaneously undermined security and, in fact, induced repeated cycles of catastrophe and trauma, making the false sense of security increasingly difficult to renounce.

The key to being labeled "civilized" in this Cartesian framework is taking control via constant monitoring. This pattern is evident from the beginning of life. For example, the dominant Western approach to educational architecture is described by Upitis (2010) as follows:

> In factory-like schools, teaching—like the buildings—tends to be boxlike and linear. Transmission teaching dominates: standing at the front of the classroom, the teacher transmits knowledge to the students who are sitting quietly facing the teacher. This type of teaching works best when the teacher can see everyone at once, and so classrooms are created as a series of boxes, the most pervasive example being one of double-loaded classrooms down a single long hallway. (p. 9)

Compare the (efficient) command-and-control approach described by Upitis with an example of an alternative (non-Western) approach provided by Swentzell (1997):

> The form and organization of the pueblo house reinforced the sense of serenity and importance of place. One sat on and played on the center of the world (the *nansipu*) and thereby derived a sense of significance. Houses were climbed on, jumped on, slept on, and cooked on. They were not material symbols of wealth but were rather, in Thoreau's terminology, a most direct and elegantly simple expression of meeting the human need for shelter ....

> Ideas that characterize the pueblo human-made and natural environments, then, are that humans and nature are inseparable, that human environments emulate and reflect the cosmos, that creative energy flows through the natural environment (of which every aspect, including rocks, trees, clouds, and people, is alive), and that aesthetics and the cosmos are synonymous (pp. 58–61).

When Europeans arrived on the shores of Turtle Island, their compulsive instinct was to take control of the landscape since this same tool set had already taken control of them. If anything, the compulsion became more urgent, as Glacken (1967) makes clear:

> Even the idea of man as a modifier of his environment acquired a dramatic character in the new lands. Men could see with their own eyes the changes—some of them temporary it is true—that they could make with fire and clearing in what many considered virgin lands unchanged since the creation. They could apply experiences and theories which had emerged in scattered localities in Europe regarding the effects of clearing and drainage. Later on, especially in the eighteenth century, the literature on man's changes in the New World environment began to grow, and men could realize both their power to change the earth and the value of apparently primeval landscapes as outside laboratories for the pursuit of nature's secrets. (p. 358)

The legal concept of *Terra Nullius* held that these "new" lands were empty of civilized peoples, (as defined by the Western command-and-control imaginary) and therefore available to be worked and improved (drained and cultivated with non-indigenous species) in accordance with directives received from God. The "wild natives" lived without recognizing the obvious-to-Europeans separation between the mastered and the un-mastered; they did not suffer from Cartesian hallucinations. They lived in what was conceptualized by the colonists as a "state of nature," meaning without (Western) civilization and, by extension, incapable of exercising mastery over the land and, especially, incapable of owning property. Thus, on one side of the equation, we have the dominators, the authors of instructions and, in consequence, the God-mandated authors of worlds, and on the other we have the inverse, the submitters-the receivers of instructions, raw materials to be molded and formed to the will of the civilized.

Of course, Indigenous peoples did modify environments significantly, but that modification process was not generally informed by an epistemology of anxiety. According to Hardess and Fortier (2014),

> Long-term planning is not new to First Nations; people don't survive in an often harsh country like Canada without the ability to plan ahead. The rich culture seen

> across the lands and waters is evidence that First Nations in Canada did in fact thrive and dedicate time to art, music, and dance. And in many instances these "cultural events" were closely tied to both short-term and long-term sustainability of the nation. (pp. 144–145)

This difference in orientation between the Western-Cartesian orientation and Indigenous epistemologies is locatable in the dominating role of trauma, fear, and anxiety in the former and the developing tools of material engagement this has spawned. Richter (2001), for example, underscores the differing orientations characteristic of European and Indigenous weapons:

> As tools for hunting, firearms were hardly the most important new imports. Early seventeenth-century matchlock muskets were so heavy that they had to be propped up on a forked support to be fired, and they were virtually useless in wet weather, which tended to extinguish the smoldering length of fuse used to ignite the gunpowder. That ignition produced a loud report prior to a still louder blast that expelled the ball from the barrel; swift prey such as deer or birds got a split second's warning to flee, and the hunter had no chance of a second shot, which could occur only after an awkward reloading of the weapon through its muzzle. Finally, if a ball did make contact, it was likely to do severe damage to a valuable pelt or hide. All told, for hunting, the bow and arrow remained more predictably accurate, stealthily quiet, quickly reloadable, and reliably lethal. (pp. 46–47)

What was it then about the musket that determined its adoption and persistence? This is not simply a case of being swept up with the sheer novelty of the tool. Europeans invaded a world full of novelty, Turtle Island was in many ways unlike anything they had previously experienced; yet they were, for the most part, irrationally loath to adopt Indigenous technologies. There must have been something about the mediational means in relation to the musket that brought forth a meaning that resonated with the Cartesian myth-system. Perhaps the musket symbolized separation and mastery over nature, in the forms of the material used and the "unnatural" results of the musket's operation—the loud bang, the smoke, and so forth.

This template of dominant versus submissive, or the sorting of everything into either an instruction sender or an instruction receiver, informs nearly every aspect of the Western way of being in the world. Although its genesis may lie in knapping hand axes, several wrong turns were still required after this initial alteration to the human mind. Conceptualizing the hand axe within a utilitarian frame, where the tool is simply useful to the conceptualized-as-separate master, as opposed to a constitutive frame, where the tool co-constitutes a person-tool assemblage, might have been the first false step.

Once this assumption of localized and monopolized agency was internalized as a dominant pattern, it spread throughout the other cultural/cognitive tools, material artifacts, and environments that wove together a social mind antithetical to sustainability.

Cartesian-mediated peoples do not necessarily regard themselves as hiding and cowering inside sterilized fortresses agonizing over violence, pestilence, or disease. But if we consider the material culture, the shared narratives, and the aesthetic values of Western-Cartesian peoples, the fear informing the design of mediational means becomes discernable. An underlying tone of fear continues to organize even the most mundane aspects of the Western-Cartesian way of becoming in the world.

Bodies

Western "development" continues to transform body-brain-place into increasingly fragmented, individuated, and disconnected collections of materialized unsustainability. For example, bodies in North America have been subjected over the past century to regimens of overprescribed antibiotics, the consequences of which we are just now beginning to understand with new research on the importance of gut fauna for human health and normal brain development (Schmidt, 2015). DDT, Roundup, thalidomide, and many other increasingly exotic cocktails of chemicals and drugs, both industrial and personal, pharmaceutical and illicit, have also terrorized bodies. Drugs, whether administered with the blessing of modern medicine or used illicitly to self-medicate in an attempt to escape sociocultural anxiety or increasingly mundane forms of terror, fit into the dominant eradication of contagion narrative. Bodies have also been terrorized by increasingly pestiferous landscapes as the entire planetary ecology has come under systematic assault by the Western social mind (Shiva, 2013). Even the accelerating loss of biodiversity is felt in the body. We might even say that the erasure of psychologically beneficial biodiversity in the form of species from passenger pigeons (1914) to heath hens (1932), Carolina parakeets (1920), Black Cisco (1960), and Blue Pike (1970) constitutes a type of ecological terrorism. Recent research suggests that healthy psychological development is greatly enhanced by time spent in natural or naturalized places. These places and their psychologically beneficial effects depend largely on biodiversity.

We should not doubt that such damage to the tissues of life impacts people at a visceral and corporeal level. Consider research on the role natural (non-anthropogenic) sounds play in cognitive performance and attention restoration (Abbott, Taff, Newman, Benfield, & Mowen, 2016) or how "natural" landscapes (meaning outside the processes of the Western brain-body-place triad) facilitate stress recovery (Ulrich et al., 1991) or recovery times from surgeries (Ulrich, 1984). To assume that our relations with other species are somehow less important to the health of children or to the social relations of parents is to commit a deep cultural/ontological error (see also Kohn, 2013; LaDuke, 1999). Healthy psychological development is not something that simply happens inside brains, but rather involves building connectivity across the many domains of environment—through brains and bodies of both humans and nonhumans and throughout the physical, architectural, material, cultural, and spiritual worlds.

The enabling cognitive tools of disinterestedness, ownership, control, and so on, get lodged in bodies, minds, and materiality. My own attempt to encourage a naturalized garden (see Chapter 6 and Figures 6.1–6.4) led to a "notice of infraction" from the city that continues to terrorize, even after the city has changed their bylaw and decided to encourage others to follow our example. I awoke this morning from a nightmare in which our entire garden, except for the large spruce that was there when we arrived 18 years ago, had been razed to the ground. The garden had been rolled in bundles and secured with baler twine and was being loaded on a logging truck along with two massive, felled, sugar-maple logs, each over a meter in diameter and over 10 meters in length. My initial thought was what had become of the young rabbits that had recently survived the raids of crows who had eaten their siblings. In the dream, I noticed no trace of the rabbits, neither fur nor blood anywhere. It occurs to me now that there would surely have been traces of carnage had I searched our real garden in the aftermath of such a horror. Instead I went, in the dream, to meet the author of the destruction and to inquire about the rabbits' fate. I went dispassionately, with an air of disinterestedness, as a proprietor would exercise rightful control over one's assets. The truck driver could be of no assistance. He was just "doing his job."

When I reflect on the dream, I wonder about the meaning of the two logs. Were they symbols of a past erasure from the Island of Montreal of trees that could have produced such logs and/or of an uncertain future when such trees might one day return? Such nightmares tap into bodily traces of mundane trauma running through synapse and sinew.

How though do such traces, emerging from the reflexive enactment of the neoliberal social mind, remain undetected, undiagnosed, and untreated? In conjunction with neoliberal logic and materiality, the operation of mind becomes an exercise of constantly knitting traumatic ruptures back into the fabric of life. New and emerging knowledges about the treatment of post-traumatic stress disorders (PTSD) with virtual reality prescriptions are illustrative (Väliaho, 2014). The basic procedure involves identifying the initial event that triggered the psychological rupture, recreating the sense of the place where the trauma was initially experienced within a "safe" virtual-reality experience, and making that experience mundane through repetition. As Väliaho explains:

> Here the correspondence between the immunitary logic of self-preservation and the therapeutic logic of the virtualization of wars, on the one hand, and the political and economic logic of contemporary wars, on the other, becomes apparent. Just as, in the epistemo-politics of PTSD, virtual reality therapies such as Virtual Iraq see the psyche as a war zone, where defense against threats to it becomes priority number one, and where images are evoked for the sake of self-preservation, to fight against pathogenic memories and to make patients operative and flexible again, so Western societies see the globe as a war zone, where they fight to defend and expand the neoliberal way of life against threats to it, incorporating the outside under their rule. Thus, virtual reality images, on one level, and current wars, on another, both positively promote the neoliberal way of life and its impulse to adapt and expand, incorporating the outside to the point where inside and outside become indistinguishable, Thus, too, "self defense" in the preemptive "war on terror" comes to mean expansion through destruction of all threats to the collective organism. (p. 86)

In other words, the contaminated psyche of the soldier suffering from PTSD is, like the threatening wilderness, in need of control. How long will it take before the errant "environmentalist" will also be explicitly, as well as subconsciously, subjected to such virtual-reality remediation? Although virtual reality may exemplify a kind of hypercontrol over body-brain dyads, real reality is not far behind. Almost any contemporary materialized cultural tool, from swimming pool to toilet, and from spacecraft to MRI scanner, can testify to this. The swimming pool and toilet are all about improving on what "wilderness" makes available to bodies. The whole Cartesian point is to *begin* by thinking the world and then for that thought to be imposed onto the world. Sewers and bodily circulatory systems merge into one process through the exercise of a will to take control of, and domesticate, fluids. These forms of material engagement know no bounds, from the deepest bodily cavities to farthest reaches of space.

Tools

Clearly, then, the tools engaged in constituting unsustainability minds (whether virtual reality, commodified interpersonal relations, or unquestioned adherence to the principles of maximizing property values) do more than act as instrumental means for achieving specified ends. Rather, through repetitive practice, the tools of the unsustainability mind routinely knit together the terrible and the insane into a "normal" that enables and perpetuates the deep dysfunction of the unsustainability mind. Benedict (1934/1989) could not have imagined how prophetic she was being in arguing nearly a century ago that "it is clear that culture may value and make socially available even highly unstable human types" (p. 170). The social-media-and-device-person, as much as the gambling addict-and-video-lottery-terminal, for example, seems to be designed as an "unstable human type" who systematically engages in a psychological retreat from the ecological crisis of our time. Similarly, the Internet netherworld has become a tool where an alternative identity, pliably programmable and easily edited and maintained to maximize presentation value, may provide a false psychological connectivity that allows a meta-anxiety to continue at the risk of triggering mass psychosis. The redirection of efforts, from controlling "wilderness" to controlling the inner world, may mark a tactical retreat as "nature" becomes increasingly unruly.

From Frankenstein's monster and Moby Dick to Kurzweil's (1999) monstrous vision of directed human-machine metamorphosis, popular culture has become increasingly saturated with the cautionary tales and dysfunctional tools of escapism. After billions of years of zigzag evolutionary processes, of coming into increasing correspondence with the outwardly infinite variety of Earth systems, seemingly sane and rational grown adults are now working on the prospect of transplanting the Earth-evolved humans to other planets. Those planets will be missing the Earth fauna and flora that live on our skin and in our guts, missing the "dust" human bodies emerge from and return to. They will be missing the gravitational forces of the Earth and the psychological support of grasses, flowers, shrubs, and trees living and self-propagating within symbiotic systems. Yet the tools of Western science provide the otherwise missing link in the Western loops of increasing anxiety needed to keep the charade going. The sociocultural anxiety born of Western ecological incompetence generates strange tools indeed.

Places

Interest in the model of *refugia* has, in large measure, been sparked by sociocultural-anxiety-driven ecological incompetence spreading throughout the environment from nanoparticle to planetary scales. Refugia are in some sense an indicator or index of ecological danger and often represent a mediated response grounded in dualism to the specter of collapsing biodiversity that stalks an ever-increasing proportion of life-forms, although this term was not originally applied to anthropogenic dangers. "The concept of identifying places of climate refuge for vulnerable species, known as *refugia*, dates back to Quaternary paleoecological studies that tried to pinpoint the locations in which species found refuge during the ice ages of the last two million years" (Hill, 2013, p. 192).

Unfortunately, within the regime of economism, the connectivity required for refugia to become more than simply a brief stopover on the journey to extinction becomes financially unjustifiable. Their continuing existence is always in opposition to the dictates of market logic, but within this context, even refugia can become places of terror. For example, in the zoo or other *ex situ* biodiversity failsafe, the preservation of genes and genomes, rather than organism-in-habitat or species-in-world, makes sense. They become materializations of an atomism where pieces are more real than wholes. Yet Haraway (2016) recognizes that "right now the earth is full of refugees, human and not, without refuge." She suggests that the "Anthropocene marks severe discontinuities; what comes after will not be like what came before .... Our job is to make the Anthropocene as short/thin as possible and to cultivate with each other in every way imaginable epochs to come that can replenish refuge" (p.100). The question remains whether this is possible when it involves swimming against such a strong cultural current.

The wheels have already come off the wagon of mainstream economic theory and even the idea of refuge may not survive the ways that place, and economistic narratives continue to become inextricably intertwined. Refugia must stay where the market puts them, which unfortunately is not where they might best function to serve the world's refugees. However, the ecological devastation inherent in economism creates an obvious need to *do something*.

Refugia are therefore politically and conceptually enclosed, their borders sealed in the belief that there is no alternative (TINA). Triage protocols in the fight to "save" biodiversity from extinction continue to be established, the logical extension of a too-obvious failure of the "biodiversity market" to allocate resources in a manner consistent with long-term human (and non-human) survival (see Brockington & Duffy, 2010; Hrabanski, 2015). Refugia, in other words, becomes something of a terrible hallucination induced by the marketing mind, exemplifying the pattern of devastation and exchange imposed on the land by the logic of the market. Contemporary refugia represent an attempt to solve a King Midas type of problem—the systematic transformation of the world into capitalist paradise and techno-Eden simultaneously makes the world uninhabitable. The paradoxical need for and impossibility of refugia turns the neoliberal world inside out, making the horror of the contemporary Self-destructiveness clear.

Narratives

The marketing mind's imperial adventures are not limited to the realm of physical geography. Even the Ivory Tower is under assault. A recent online call for papers for an international conference on "Environmental Science and Development" provided the following list of suggested topics (ICESD, 2018):

Environmental Science and Technology
Environmental dynamics
Meteorology
Hydrology
Geophysics
Atmospheric physics
Physical oceanography
Global environmental change and ecosystems management
Climate and climatic changes
Global warming
Ozone layer depletion
Carbon capture and storage
Biofuels
Integrated ecosystems management
Satellite applications in the environment
Environmental restoration and ecological engineering

Habitat reconstruction
Biodiversity conservation
Deforestation
Wetlands
Landscape degradation and restoration
Ground water remediation
Soil decontamination
Eco-technology
Bio-engineering
Environmental sustainability
Resource management
Life cycle analysis
Environmental systems approach
Renewable sources of energy-energy savings
Clean technologies
Sustainable cities
Health and the Environment
Health-related organisms
Hazardous substances and detection techniques
Biodegradation of hazardous substances
Toxicity assessment and epidemiological studies
Quality guidelines, environmental regulation, and monitoring
Indoor air pollution
Water resources and river basin management
Regulatory practice Water quality objectives standard setting
Water quality classification

And so on, for a further 40 similar topics. The list creates a narrative just as the real-estate section of the classified advertisements in any local newspaper does: by framing the object in the most impoverished way conceivable. Real-estate advertisements use the minimum number of words possible to convey only information that might impact the market price of a given location. They generally omit descriptions of the historic, geological, hydrological, ecological, or phenomenological characteristics of the place. Calls for papers on "environmental science and development" that ignore the workings of the human social mind work similarly to frame the world as one clearly demarcated into mutually exclusive categories, each of which suggests the need for further knowledge and control. There is no acknowledgment on this list of any of the circuits I have been describing running between places and human systems of meaning, just a profound form of cognitive and emotional disconnection. The story it tells is one of a debased world, only comprehensible through the most impoverished of metrics.

When sustainability is reduced to "management of hazardous solid waste" or "water treatment and reclamation," we accept a model of engagement that incarcerates us within the practices of unsustainability. How, exactly, is a world thusly storied, embodied, materialized, and practiced any different from the storying, embodying, materializing, and practice of the Romanian orphanage? What kind of life is possible within its emotionally, phenomenologically, and ethically stripped framework?

Today tens, if not hundreds, of thousands of people live on the fringes of vast garbage dumps in extremely precarious conditions, gathering whatever bits of matter can be used, sold, bartered, or eaten (see Kamel, 2002). The story and storytelling of such peoples are conspicuously absent from the frames or metaphors found in this call for papers, just as they are in real-estate flyers. These people are not easily captured by any boundary lines drawn between domesticated and feral. Their stories, their potential happiness, and their psychological health rely more on frames and metaphors that promote connectivity and cooperation than ones that promote competition, distinct boundaries, and accumulation. The stories told from the perspective of capitalist enterprise regard the categories of garbage, industrial waste, and surplus labor as unfortunately inevitable but also easily ignorable. But garbage is in no sense ignorable for these thousands who depend upon it. To tell their stories through the categories suggested in the call for papers would do an injustice to their suffering; it would disconnect us and dehumanize them (and vice versa). Such approaches clash with understandings of a world that is profoundly interconnected. The same might be said of the reduction of other places that tell stories of terror—tar-sands tailings ponds, vast oceanic gyres of plastic, and an atmosphere and a hydrosphere filled with industrial pollution warming the planet, acidifying the oceans, and inflicting kinds of chemically induced suffering previously unimaginable. To tell these stories in the categories approved for academic scientific papers or real-estate flyers is to obscure the need for us to ever end that worldly suffering.

Processes

Even from birth, the process of forcing children to fit the dominant narrative begins. For example, small children can sometimes experience separation anxiety due to the absence of the significant caregiver or a transitional object such a blanket or teddy bear (Oliveira Dias, 2016, pp. 118–125). In a world of disconnection and atomistic tools, the transitional object, not quite self and not quite other, can ease the process of being mastered by Cartesian mediational means that determine that the developing child must become entirely separate. While Western society generally sees these anxieties as developmental blips in the inexorable process of psychological development on the path to self-mastery, I am inclined to wonder whether these tendencies, more powerful in some children than others, might reflect at some level an innate model for the subjective experience of sustainability. Perhaps discouraging extensions of the self, whether by separating the child from transitional objects or stopping the child from sleeping in the same bed with parents, is linked with conceptualizing the world as simply lifeless and malleable matter, and thus with transforming everything into inert "resources" to be mastered and reconfigured according to the dictates of market forces. It is doubtful whether separation anxiety could be much of an issue in a sociocultural world in which the world was thought to be alive and, for the most part, friendly.

The central focus of this manuscript has been the processes whereby persons are woven together with places to form patterns of (un-) sustainability, and a key aspect of this process is parenting and the quality of the parental environment. In Winnicott's terms, healthy child development requires "good enough mothering" (see Modell, 1968 p. 87), but this seems to assume that parenting is just something that occurs between parent and child. But, from the position I have been arguing, "goodness of parenting" should also be considered as a descriptor of the physical places we inhabit and the material culture that structures our action. In a world of extended and distributed cognition, where "people crystallize out in the interstices between objects, taking up space allowed them by the object world, with our sense and emotions educated by the object world" (Gosden, 2005, p. 197), it is important to keep in mind that environment, taken more generally, may not meet the criteria of "good enough" parenting.

Our current understandings of the environment that parents us, the biosphere, seems to suggest that its processes are organized according to principles of competition. Such expectations are also embedded in the instruments, lenses, preformed conceptual categories, and other tools we use for seeking, seeing, feeling, understanding, and experiencing that environment. This, of course, is hardly likely to provide a "good enough" model for parenting. Learning to compete with a world, to always be taming and mastering, is part of the Western historical inheritance that acquires us from birth. But with a different set of mediational means, a different world would become the dominant process of enactment. From a different perspective, a different cultural lens, the world could be conceptualized as constitutive of the self and organized around the principle of love. We would then be woven into a world through a developmental process not of fear, anxiety, and terror, or systematic impoverishment, but rather through processes of mutuality, love, and sustainability.

Agency

Where is the agency that drives the sociocultural anxiety of the Western imaginary to be found? This is a question for the historian, the historiographer, the psycho-historian, the cognitive archaeologist, and the forensic ethnographer. Mistakes of our ancestors—past mistakes of thought, feeling, action, and institutionalized structures and material artifacts—clearly have extreme consequences in the present, and yet our contemporary sociocultural configurations systematically deny their relevance. We understand ourselves as a ghost-in-machine operating on a tabula rasa and fall under the hallucinatory spell of a cultural historical inheritance that suggests we each have a monopoly position with respect to agency. In truth, we exercise very little control, especially where the power to change the cognitive tools structuring our thoughts and feelings are concerned (see also Chapter 1). We all become hostages to a past that focuses on control as the antidote to, rather than the result of, a zigzag process of ceding agency to materialized anxiety. We are, as Foucault (2008) argued in his discussion of the organization and policing of space during a plague outbreak, disciplined by mechanisms that exceed individual agency. This disciplinary mechanism

> lays down for each individual his place, his body, his disease and his death, his well-being, by means of an omnipresent and omniscient power that subdivides itself in a regular, uninterrupted way even to the ultimate determination of the individual, of what characterizes him, of what belongs to him, of what happens to him. (p. 3)

In this history of agency, power was increasingly vested in the mediational means generated by the plague and was not simply a matter of people consciously willing to barter their agency away in return for psychological security.

The zigzag nature of the way world works, as outlined by Bateson (1972/2000), suggests that the agency of anxiety is both distributed and self-reinforcing.

> Unaided consciousness must always tend toward hate, not only because it is good common sense to exterminate the other fellow, but for the more profound reason that, seeing only arcs of circuits, the individual is continually surprised and necessarily angered when his hardheaded policies return to plague the inventor.
>
> If you use DDT to kill insects, you may succeed in reducing the insect population so far that the insectivores will starve. You will then have to use more DDT than before to kill the insects which the birds no longer eat. More probably, you will kill off the birds in the first round when they eat the poisoned insects. If the DDT kills off the dogs, you will have more police to keep down the burglars. The burglars will become better armed and more cunning … and so on.
>
> That is the sort of world we live in—a world of circuit structures—and love can survive only if wisdom (i.e., a sense or recognition of the fact of circuitry) has an effective voice. (p. 146)

Of course, "unaided consciousness" only exists as a theoretical construct within the Western-Cartesian worldview. It is a map without a corresponding reality. The point, though, is that person-unaided-consciousness-map-assemblages operate as within a pattern of sociocultural anxiety, a pattern of self-destruction. It is then "the fact of circuitry," that Bateson rightly identifies, combined with the mediational means generated by the unskillful operation of that circuitry, that becomes the primary agent "causing" the mischief.

Memories

In North America, as in other places still struggling to come to terms with their colonial pasts, there has been a systematic erasure and forgetting of the pre-existing Turtle Island persons-tools-environment (mind) assemblage. The Turtle Island landscape was providing "good enough parenting" so that Indigenous peoples largely avoided ways of becoming that entailed the destruction of the land's best affordances. They avoided plagues, pandemics, and the ecocides that formed the foundations of the Western gestalt. Richter (2001), incorporating a quote from Carolyn Merchant's (1989) *Ecological Revolutions*, explains:

> The arrival of European farmers—with their roaming livestock, their concepts of fixed property, and their single crop agriculture—combined with the ecological impact of the fur trade to transform utterly the material environment of much of eastern North America and make traditional patterns of life impossible anywhere in the vicinity of European settlements. European and Indian ways of using the land could no more share the same ecosystem than could matter and antimatter share the same space. "Our fathers had plenty of deer and skins, our plains were full of deer, as also our woods, and of turkeys, and our coves full of fish and fowl," the Narragansett headman Miantonomo concluded in 1642. "But these English having gotten our land, they with scythes cut down the grass, and with axes fell the trees; their cows and horses eat the grass, and their hogs spoil our clam banks, and we shall all be starved." (p. 59)

The Western mind cycling of induced ecological impoverishment, trauma and sociocultural anxiety—all conventionally understood at every stage as an indicator of continuing progress—means that only traces of the memory of this previous way of life, and means of becoming, still exist. Some of us would like to imagine that the land can remember what the incoming people chose to forget, that the land, too, is a repository for memory. This is clearly true, but questions remain as to what the land has also forgotten, what memories have been erased, what land could and perhaps should forget of colonial interventions, and what memories might be recalled in the present. For example, some animals and plants have adapted quite well to the Western-Cartesian changes in the land. Others have not and some are now remembered only in the geologically ephemeral cultural tools of peoples. Nobody pays much attention to long-term trends in the life expectancies of trees or what such changes mean for various soil organisms and other species dependent on older trees, for example. The land must find a way to carry on regardless. That does not diminish in any way the ecological horror, but it is important not to enable, encourage, or reinforce the memories that perpetuate that horror.

We might likewise imagine the horror of each and every future childhood impoverished by the extinction of frogs and salamanders, a prospect now hurtling closer to reality with every passing hour. But again, questions remain as to whether focusing on the horror of these extinctions helps us extricate ourselves from the mediational means responsible for such losses or just perpetuates and amplifies the cycle of sociocultural anxiety and unsustainability. Burkell (2016) explains why we should, at times, be mindful of the memories we hold on to:

> Lived experience reinforces the notion that memory is critical to identity: for example, conditions that lead to impairment in autobiographical memory (e.g., Alzheimer's

> disease) are said to lead to "loss of self." Faced with the specter of a self that disappears as life memories erode, it is tempting to conclude that a flawless biographical memory is the optimal state. Research on hyperthymesia, or "unusual autobiographical memory," however, suggests otherwise, and those living with the condition describe their memories as "tyrannical." By virtue of their inescapable accuracy and completeness, these memories compromise the ability to construct and recraft integrated personal narratives that tell us, and others, "who we are." (p. 18)

Cultural memories, especially when marinated in sociocultural anxiety can become "tyrannical." In the matrix of twenty-first-century unsustainability, the arts of memory must be practiced skillfully, without succumbing to the dominating tone of fear, anxiety, and ecological terror and, especially, without commemorating a destructive past. Disneyland and shopping malls provide two examples of an approach to memorializing something that never really existed, another suppression-of-anxiety technique. Skillful remembering is antithetical to this type of escapism. Skillfully recovering ecological memories will involve attention to processes and relationships as much or more than attention to species. It must be further recognized that those processes and relationships are as much mental as they are material.

For the most part, this is not currently what we find. Many of the place names where I now live are those of mainly white, male, French or British settlers. Far fewer reflect aspects of physical environment. Those that do, unfortunately, often recall a material environment that no longer exists. *Orme* (French for elm) is a common root for many local place names such as Ormsetown, l'Anse à l'Orme or Cap à l'Orme. Yet, the elm trees were killed off by Dutch elm disease, and today's inhabitants, disconnected from nature, have no trouble forgetting the tree that once dominated the landscape (soon to be followed by the ash trees dying from the emerald ash borer). The very few Indigenous place names that still act as reminders generally designate, and are contained within, the reserves that act as a kind of cultural refugia.

Atmospheres

Many might find the idea that modernity is underlain with a deep subconscious psychic terror problematic. But surely this becomes more apparent if we examine the links between Western anxiety and global terror. The War on Terror seems the perfect example of a Batesonian double bind, and fear explicitly underlies both those attacking the salient symbols of capitalism such as the World Trade Towers via acts of terrorism and those imposing campaigns of "shock and awe" by bombing the Middle East in the name of Western values. Each society draws upon a distinct repertoire to make sense of the transformative experiences of capitalism and to "organize a response in ways which connect with defining values and symbols of [their respective] inheritance" (Tripp, 2006 pp. 200–201). The fear embedded in capitalism projects, distributes, and enforces the requisite atmospheres of marketing anxiety even to those cultures opposed to it. The fear of capitalism and the fears produced by capitalism are part of the collective creating and maintaining atmospheres of anxiety and terror around the respective objects of fear. The practitioners of terrorism detect and exploit by weaponizing the Western fear, while at the same time realizing and reacting to their own fears of becoming Westernized.

The Western-Cartesian mediational means are likewise organized in opposition to that which is perceived and understood as Other. The white supremacist fears different peoples whom they have not tried to understand and, in consequence, foreclose the potential development of empathy and connectivity to a broader community. But these same fears, so easily recognizable as abhorrent in the white supremacist, also run like a thread throughout the fabric of North American society and include both "othered" humans and "othered" nature and with similarly disastrous and violent consequences. Again, this might be linked back to an atmosphere of fear that transcends political views and pervades Western understandings of nature. To take just one example, in 1874, John Stuart Mill (1998) had this to say on the topic of "Nature":

> The scheme of Nature regarded in its whole extent, cannot have had, for its sole or even principal object, the good of human or other sentient beings. What good it brings to them, is mostly the result of their own exertions. Whatsoever, in nature, gives indication of beneficent design, proves this beneficence to be armed only with limited power; and the duty of man is to co-operate with the beneficent powers, not by imitating but by perpetually striving to amend the course of nature—and bringing that part of it over which we can exercise control, more nearly into conformity with a high standard of justice and goodness. (p. 65)

Mill's scathing assessment of "Nature's" ability to parent is striking and the deep, subconscious anxiety that informs the agonistic atmosphere between the civilizing forces of the civilized against the savagely unknowable wilderness lies at the root of these anxieties (Bauman, 2017).

This atmosphere of anxiety forms the context of psychological development for every child born into the Western mind. From the inside, no alternative seems to be possible, but from the outside, capitalism appears as a truly impoverished and fearful way of being in the world. It lacks connectivity; in fact, it actively works to suppress the kind of connectivity that would refute the notion that things could be anything more than simple commodities. Capitalism works to undo empathy and the kind of intersubjectivity that could enable right and just relations with the world or the development of workable ethics, environmental regulations, or a sane foreign policy. Capitalism leaves people lost and afraid, and that fear and disorientation feeds back into a double bind that ensures support for, and continuation of, capitalism.

Sustainability

The implicit anxiety and the resultant terror of the Western-Cartesian way of being in the world act as dominant organizing forces working to weave peoples into a shared praxis and habitus that is antithetical to sustainability. In some ways, this is similar to the kinds of shared sensibilities that Berleant (2014) regards as necessary to explain changing aesthetic predilections.

> By sensibility I mean perceptual awareness that is developed, focused, and informed. It is more than simple sensation, more than sense perception. Perhaps one can consider it educated sensation. It requires the perceptual knowledge and skills that we are continually enhancing in and through our encounters and activities. Aesthetic sensibility develops and uses this capacity at the deliberate center of conscious experience. In Western cultures, the arts have been the primary medium for promoting such awareness, and we can consider changes in artistic style, the emergence of new movements, and even entire historical periods in the arts as fundamentally changes in sensibility. Looking at culture change more broadly, we can regard fashion, etiquette, and behaviour patterns in general as expressions of the prevailing sensibility of a place and time. (p. 19)

The sensibility of (un-) sustainability is, in other words, a *mediated* sensibility. And while Berleant argues that "the arts ... fashion, etiquette, and behaviour patterns" are indexical of an acquired and shared sensibility, we might add that so, too, are toilets, transcontinental ballistic missiles, teapots, and teepees. The mediational means of (un-) sustainability arise from the ways the entire world has been configured by those who came before us. As this previous the chapter have argued, two of the dominant themes or orientations of our contemporary unsustainability sensibility, the conjoined twins of economism and anxiety, invoke a sensibility that is antithetical to sustainability. The next chapter will illustrate the (un-) sustainability consequences of that sensibility, how it relates to various forms of material engagement mediated by capitalism and how it comes to be materialized in the (aesthetic) form of mowed lawns.

References

Abbott, L. C., Taff, D., Newman, P., Benfield, J. A., & Mowen, A. J. (2016). The influence of natural sounds on attention restoration. *Journal of Park and Recreation Administration*, *34*(3), 5–15.

Bateson, G. ([1973] 2000). *Steps to an ecology of mind*. Chicago, IL: University of Chicago Press. (Original work published 1972).

Bauman, Z. (2017). Symptoms in search of an object and a name. In H. Geiselberger (Ed.), *The great regression* (pp. 13–25). Cambridge, UK and Malden, MA: Polity.

Benedict, R. (1989). *Patterns of culture*. Boston, MA: Houghton Mifflin. (Original work published 1934).

Berleant, A. (2014). Environmental sensibility. *Studia Phaenomenologica*, *14*, 17–23.

Brockington, D., & Duffy, R. (2010). Capitalism and conservation: The production and reproduction of biodiversity conservation. *Antipode*, *42*, 469–484.

Burkell, J. A. (2016). Remembering me: Big data, individual identity, and the psychological necessity of forgetting. *Ethics & Information Technology*, *18*(1), 17–23.

Byers, L., Kulitja, S., Lowell, A., & Kruska, S. (2012). "Hear our stories": Child-rearing practices of a remote Australian aboriginal community. *Australian Journal of Rural Health*, *20*, 293–297.

Clark, M.E. (2002). In search of human nature. London, UK and New York, NY: Routledge.

Cosgrove, D. (2001). *Apollo's eye: A cartographic genealogy of the Earth in the Western imagination*. Baltimore, MD: John Hopkins Press.

Eliot, J. (1748). *An essay upon field-husbandry in New England*. New London, CT: T. Green.

Flavell, J. (1709). *Husbandry spiritualized or the heavenly use of earthly things* (10th ed.). Boston, MA: John Allen.

Foucault, M. (2008). "Panopticism" from d*iscipline & punish: The birth of the prison*. *Race/Ethnicity: Multidisciplinary Global Perspectives*, *2*(1), 1–12.

Gifford, E., & Gifford, R. (2016). The largely unacknowledged impact of climate change on mental health. *Bulletin of the Atomic Scientists*, *72*, 292–297.

Glacken, C. (1967). *Traces on a Rhodian shore: Nature and culture in Western thought from ancient times to the end of the eighteenth century*. Berkeley: University of California Press.

Gosden, C. (2005). What do objects want? *Journal of Archaeological Method and Theory*, *12*, 193–210.

Haraway, D. (2016). *Staying with the trouble: Making kin in the Chthulucen*. Durham, NC: Duke University Press.

Hardess, L., & Fortier, K. J. (2014). Co-creative planning: Simpcw First Nation and the Centre for Indigenous Environmental Resources. In R. Walker, T. Jojola, & D. Natcher

(Eds.), *Reclaiming indigenous planning* (pp. 166–192). Montreal, QC: McGill-Queen's University Press.

Harrison, R. P. (1992). *Forests: The shadow of civilization*. Chicago, IL: University of Chicago Press.

Hill, K. (2013). Biodiversity and climate change in cities. In J. Beardsley (Ed.), *Designing wildlife habitats*, pp. 187–206. Washington, DC: Dumbarton Oaks.

Hinchliffe, S., Bingham, N., Allen, J., & Carter, S. (2017). *Pathological lives: Disease, space and biopolitics*. Malden, MA: Wiley.

Hrabanski, M. (2015). The biodiversity offsets as market-based instruments in global governance: Origins, success and controversies. *Ecosystem Services*, *15*, 143–151.

Hundert, E. M. (1997). *Lessons from an optical illusion: On nature and nurture, knowledge and values*. Cambridge, MA: Harvard University Press.

International Conference on Environmental Science and Development (ICESD). (2018). *ICESD 2018 Call for papers*. Retrieved from http://www.icesd.org/topics.html

Kamel, L. I. (2002). Ecological education in the living environment. In J.-P. Hautecoeur (Ed.), *Ecological education in everyday life* (pp. 21–43). Toronto, ON: University of Toronto Press.

King, T. (2003). *The truth about stories: A native narrative*. Toronto, ON: Anansi.

Kohn, E. (2013). *How forests think: Toward an anthropology beyond the human*. Berkeley: University of California Press.

Kurzweil, R. (1999). *The age of spiritual machines: When computers excess human intelligence*. New York, NY: Penguin Books.

LaDuke, W. (1999). *All our relations: Native struggles for land and life*. Cambridge, MA: South End Press.

Le Moigne, J.-L. (2013). The intelligence of complexity: Do the ethical aims of research and intervention in education not lead us to a new discourse "on the study methods of our time"? *Complicity: An International Journal of Complexity and Education*, *10*, 1–17.

Louv, R. (2005). *Last child in woods: Saving our children from nature-deficit disorder*. Chapel Hill, NC: Algonquin Books.

Louv, R. (2011). *The nature principle: Reconnecting with life in a virtual age*. Chapel Hill, NC: Algonquin Books.

Malafouris, L. (2013). *How things shape the mind: A theory of material engagement*. Cambridge, MA: MIT Press.

Merchant, C. (2008). Secrets of nature: The Bacon debates revisited. *Journal of the History of Ideas*, *69*, 147–162.

Mill, J. S. (1998). *Nature, the utility of religion, and theism*. Amherst, NY: Prometheus Books. (Original work published 1874).

Modell, A. H. (1968). *Object love and reality: An introduction to a psychoanalytic theory of object relations*. New York, NY: International Universities Press.

Oliveira Dias, E. (2016). *Winnicott's theory of maturational processes*. London, UK: Routledge.

Özerdem, A., & Podder, S. (Eds.). (2011). *Child soldiers: From recruitment to reintegration*. New York, NY: Palgrave Macmillan.

Pfouts, R. W. (2002). On the need for a more complete ontology of the consumer. In E. Fullbrook (Ed.), *Intersubjectivity in economics: Agents and structures* (pp. 71–84). London, UK: Routledge.

Rhodes, R. (2012). The power of power. *History Today*, *62*(8), 37–42.

Richter, D. K. (2001). *Facing east from Indian country: A native history of early America*. Cambridge, MA: Harvard University Press.

Schmidt, C. (2015). Thinking from the gut. *Scientific American*, *312*(3), 12–15.

Shiva, V. (2013). *Making peace with the earth*. London, UK: Plato Press.

Skoll, G. R. (2010). *Social theory of fear: Terror, torture, and death in a post-capitalist world*. Basingstoke, UK: Palgrave Macmillan.

Swentzell, R. (1997). Conflicting landscape values: The Santa Clara Pueblo and Day School. In P. Groth & T. W. Bressi (Eds.), *Understanding ordinary landscapes* (pp. 56–66). New Haven, CT: Yale University Press.

Tripp, C. (2006). *Islam and the moral economy: The challenge of capitalism*. Cambridge, UK: Cambridge University Press.

Turkle, S. (2011). *Alone together: Why we expect more from technology and less from each other*. New York, NY: Basic Books.

Turkle, S. (2015). *Reclaiming conversation: The power of talk in a digital age*. New York, NY: Penguin Books.

Ulrich, R. S. (1984). View through a window may influence recovery from surgery. *Science*, *224*(4647), 420–421.

Ulrich, R. S., Simons, R. F., Losito, B. D., Fiorito, E., Miles, M. A., & Zelson, M. (1991). Stress recovery during exposure to natural and urban environments. *Journal of Environmental Psychology*, *11*, 201–230.

Upitis, R. (2010). *Raising a school: Foundations for school architecture*. Township of South Frontenac, ON: Wintergreen Studios Press.

Väliaho, P. (2014). *Biopolitical screens: Image, power and the neoliberal brain*. Cambridge, MA: MIT Press.

Vico, G. (1968). *The new science of Giambattista Vico*. Ithaca, NY: Cornell University Press. (Original work published 1744).

Vygotsky, L. V. (2012). *Thought and language: Revised and expanded edition*. Cambridge, MA: MIT Press. (Original work published 1962).

· 6 ·

THE MOWED (UN-) SUSTAINABILITY MIND

This chapter begins with a description, analysis, and discussion of that aesthetic icon of suburban material culture, the mowed lawn. This icon is then contrasted with an alternative person-tool set-environment assemblage, the naturalized, or "rewilded" (Tokarski, 2019), suburban garden. Finally, the two are compared in terms of the parameters of (un-) sustainability. The point here is not to simply to equate the mowed lawn with unsustainability and the naturalized garden with sustainability. Rather, they should be considered as exemplary settings on a continuum within a larger system of a person-tool set-environment configuration. Activities can take place on a mowed lawn that could theoretically move the person-tool set-environment configuration closer to sustainability. These activities, for example, community gatherings, recreational sports and games, or family picnics never completely erase the negative psychological and ecological impacts of the mowed lawn, but might still represent intermediate steps on the shift from unsustainability toward sustainability (Dale & Newman, 2010). The smell of freshly cut grass or grass stains on clothes might contribute to a sense of connectivity and instill a desire to maintain the viability of a community. However, at a tacit level, the mowed lawn conveys an implicit meaning that is inextricable from unsustainability and the Cartesian delusion. The mowed-lawn aesthetic exemplifies a compulsion that offers a key to understanding the wickedness of the unsustainability problem.

The process of replacing a mowed lawn with a naturalized garden can, in some sense, change those rules of the game. At the same time, a naturalized garden could conceivably be incorporated into an unsustainability ecology of mind. Much depends on the ontological model one holds in mind throughout the practice of either lawn mowing or rewilding and how that model is materialized. Consider, for example, a trip to the supermarket to buy something for supper. In the suburbs, the supermarket is about as unavoidable as the management of the front property. There are, of course, a few other options if one is willing to put in the effort to question and confront the broader capitalistic dynamo of suburban existence. One might order a weekly basket from a local vegetable grower who also sources other products from local suppliers. Taking delivery of the basket is not always convenient, but the alternative is a very convenient choice between unsustainable alternatives. The naturalized garden is similarly inconvenient when compared to the standardized routine of simply mowing the lawn. For example, if the model is centered on biodiversity, then time and effort must be expended not only in monitoring the garden for invasive alien species, but also in selecting the plants that will provide the greatest benefit from the perspective of biodiversity conservation. That same inconvenience, however, within the contemporary capitalist context, forms a sort of loose end that can potentially be woven back into the more dominant fabric of societal-scale unsustainability. Guthman (2003), for example, provides food for thought:

> Organic salad mix has come a long way from the aesthetic of the slow food gastronome, even further from the holey lettuce found at the local health food co-op. So it is striking that fast food and organic/slow food continue to be posed as binary, even organic assemblages, if you will, of taste, body type, social consciousness, class, mode of production, and so forth. Sometimes termed tendency and counter-tendency, sometimes hegemony and resistance, one of the problems with these oppositions is they impart a good deal of subjectivity on to the organic or slow food eater while the fast food eater is treated as mindless dupe. (p. 55)

Guthman then describes how a grassroots, organic-food movement in California becomes co-opted into an industrialized, corporate agriculture that then sets the rules of the game, making it impossible for local, small-scale organic producers to compete in the same markets or even to meet the requirements for being certified as organic. If the naturalized garden becomes a "fashion," it could certainly meet a similar fate, serving to reinforce what it initially reacted against.

One way this might happen is if the care of the naturalized garden followed the trajectory seen over the past decade in my hometown of Pointe Claire; here, the trend has moved away from people mowing their own lawns and toward a model where lawn-care services are provided by companies on contract. Within this model, the naturalized garden could easily become a form of green-washing, whereby homeowners advertise their differentiating ecological enlightenment at a much higher cost, but still as simply another form of conspicuous consumption. Gutham's article, then, underscores the care that must be exercised in any attempt to reunite mind and matter in a relational ontology of sustainability. New stories and new identities need to be enlisted simultaneously with any new materiality if the goal is to establish a sustainability ecology of mind.

Mowed Lawn

Mowed lawns are iconic of a peculiar type of sociocultural-ecological unsustainability. They are a materialization of the underlying economistic, traumatized, and fearful historical social psyche of the Western-Cartesian model of subjectivity. They resonate deeply with the Cartesian psyche for at least three reasons. First, the mowed lawn demonstrates a level of mastery, including both mastery of environment and mastery of self. Next, the mowed lawn is indexical of the capitalist economic system. The mowed lawn creates and sustains an atmosphere of anxiety paralleling the consumer anxiety that the market economic system depends on to maintain aggregate demand and avoid economic recession or depression (see Robbins, 2007). The mowed lawn reproduces a form of abstraction that prioritizes and sanctifies the market factors expected to assign an objective value to any specific property in question. The commodified parcel of land becomes more easily quantifiable and comparable with and to other competing parcels. Finally, the mowed lawn draws its participants into a highly organized and ritualistic forgetting, whereby the obvious insanity that the mowed lawn represents is actively covered up.

The mowed lawn requires mastery through the enforcement of an ecological regime that is nearly impossible to maintain. Because this regime is so difficult to maintain, the human enforcer must become an active participant in a larger sociocultural-ecological system of discipline and effectively becomes mastered themselves by that system. In other words, the mastery initially projected onto the land is reflexively internalized by the master, spreading control throughout the entire unsustainability mind.

This happens for a number of interconnected reasons. For example, the grass monoculture is especially susceptible to "pests." Mowing reduces the surface area of the grass leaves, and thus sets the parameters for how much photosynthesis can be performed and how much energy the plant has available for growth and sustenance. With less energy available, deep root systems cannot be supported. Shallower roots mean that fewer soil organisms can be supported below the top few centimeters of soil, meaning the entire soil ecology is impoverished through routine mowing. As the specific nutrients required by grass plants, which are not the same as those for other types of plants, are extracted and the waste products of the grass plants accumulate, weeds arrive as if invited. In a normal ecosystem with higher levels of species diversity, the by-products and wastes of some species will be the resources and nutrients for other species. Species diversity works to bring "balance" to the system. The lawn, however, becomes increasingly attractive to weed species precisely because of this increasing imbalance. The weakened state of the monocultural grass ecosystem is compounded by the fact that mowing makes the ground more accessible to sunlight and heat and therefore more attractive to egg-laying beetles and moths. With a more impoverished soil ecosystem, the ground also becomes more compacted and water retention capacity is reduced. The reduced capacity to retain water increases the likelihood that the manicured grass will turn brown. Watering the lawn, though, actually reduces grub-egg mortality, as the eggs require constant soil moisture to hatch. And so, the destructive cycle continues. In other words, the discipline of maintaining the mowed lawn invites the system to fall apart. In this sense, the practice of maintaining the mowed lawn mirrors the operation of learned models of subjectivity associated with the maintenance of a capitalist system and represents a ritualistic confirmation of the Western-Cartesian delusion.

The repetitive aspect of the lawn-maintenance regime reassuringly feeds the double-binding pattern once the investment in time, energy, financial resources, and community aesthetic norms is "sunk" subconsciously beyond recovery. Mowed lawns reinforce and continue the familiar cycles of inescapable anxiety that characterize the Western-Cartesian worldview. The charade progresses as whole communities participate in a practice that transforms them into an increasingly psychologically and physiologically toxic place to live. This process has nothing to do with a lack of scientific or propositional knowledge. It is about the implicit learning of cultural/historical scripts in the form of tacit knowledges supported and maintained within communities of practice. Mowed lawns represent exemplary nodes of unsustainability within the prevailing ecology of mind.

As already mentioned, the anxiety required to keep consumers buying stuff and thereby maintain aggregate demand is not unrelated to the anxiety that keeps homeowners mowing their lawns (Robbins, 2007). Most of what can be known and understood about mowed lawns can find a parallel in the market economic system. According to Larsen and Harlan (2006),

> The residential landscape is a repository for incompatible ideologies. Since the single-family house in the United States is private property, it embodies the Jeffersonian ideals of individual freedom, democratic participation and unconstrained opportunity. However, the reality of the residential landscape immediately challenges these ideals, which are ingrained not only in the United States, but throughout many parts of the capitalist world. Economic sorting spatially clusters "like" residents by income, and social pressures encourage conformity. Today, most homeowners view their home as an important financial investment and their desire to eventually profit from its resale also influences their attitudes and behaviors. Further complicating the meaning of the residential landscape is its connection to the "natural" landscape and, perhaps most importantly, its symbolic representation of the homeowner. (p. 86)

The mowed lawn is a very public display of human mastery over the perceived-as uncontrolled and dangerous nature. One of the residents of my community, for example, once told me that he wanted anyone looking at the manicured tidiness of their lawn to know with certainty, everything inside the house was in perfect order: clean, shiny, and exactly where it belonged. Not everyone sees the mowed lawn as an advertising medium for one's obsessive-compulsive disorder, but the mowed lawn does fit effortlessly into Simmel's (2000) "philosophy of fashion" for nearly all suburbanites:

> All feeling of shame rests upon the conspicuousness of the individual. It arises whenever a stress is laid up on the self, whenever the attention of a social circle is drawn to an individual, which at the same time is felt to be in some way inappropriate. For this reason retiring and weak natures are particularly inclined to feelings of shame. The moment they step into the centre of general attention, the moment they make themselves conspicuous in any way, a painful oscillation between emphasis upon and withdrawal of the sense of the self manifests itself. In so far as this individual conspicuousness, as the source of the feeling of shame, is quite independent of the particular content on the basis of which occurs, so one is frequently ashamed of good and noble things. (p. 199)

The practice of maintaining a mowed lawn also requires the imposition of socially generated norms that harness fear and anxiety, including practices of shaming nonconformity, to maintain order. That same obsession with control, however, undermines the sustainability of the mowed lawn eco-mental system.

Like any ritual, people take comfort in the simplified symbolic action of repetitive enactment of the underlying myth. Antibacterial soaps, overprescribed antibiotics, genetically modified crop species (GMOs), and agricultural pesticides all fit into the same recursive inheritance forged in the kilns of anxiety and an unquenchable quest for control within a context of recurring societal experiences of pestilence, famine, soil degradation, armed conflict, and ontological delusion. There is nothing unusual about this pattern of generating strange rituals. In the same way that an athlete might find comfort in an article of clothing worn during a previous successful outcome that provides proof that the game *can* be won, the manicured mowed lawn provides a comforting reassurance that savage, dangerous nature *can* be conquered and managed. Like any double bind, however, the system eventually falls apart as the psycho-ecological consequences begin to pile up.

The mowed lawn wears the soil out and requires increasingly heroic efforts to maintain the uniform green aesthetic. Fertilizers and soil supplements need to be brought in from somewhere else, and frequent irrigation may lead to salinization, thus leading to further soil amendments. Moss killer, weed killers, and pest sprays need to be deployed (all of which have further knock-on ecological and health effects). New tools and irrigation systems need to be deployed, all dictated by, and dictating, particular forms of material engagement. It is in this continuous process of collapse that the lawn ecosystem comes into a mediated subjective correspondence with the marketing mind. According to Jenkins (1994),

> The basis of America's front-lawn aesthetic is the ideal of a single type of grass that will stay green year round. With the cultivation of a large number of plants of the same type comes a greater susceptibility to disease and insect damage. The standard held up for lawns that requires the grass to be a single colour, texture, and size was unrealistic even for the most devoted gardener; however, it enabled manufacturers to sell more weapons to home owners in their battle against nature. (p. 158)

It is, in other words, the falling apart of the lawn that constitutes a marketing opportunity. The subjective experience of maintaining a mowed lawn exemplifies the subjective experience of participating in a capitalist economic system, including the underlying fear, shame, obsession with control, the assumption of a world controllable through the application of scientific methods, and so on. The mowed lawn is an icon of the Western unsustainability *modus vivendi.*

The suburban person–tool set–environment constellation, of course, includes more than simply the mowed lawn (see, for example, Baumgartner, 1988). We could discuss with equal import the pattern and planning process of suburban building construction, the reaction to pests inside suburban houses, or the tacit assumptions about mobility, social justice, or linkages between consumerism and identity. The suburban-landscaping regime should be interpreted as merely one thread in a complexly woven whole cloth. At the same time, it should be recognized that by changing keystone threads, sometimes the whole cloth can be transformed. Converting a mowed lawn to a naturalized garden (Figure 6.1) can substantially alter prosthetic cognitive and emotional processes for the participants of that changed constellation.

Figure 6.1. A Naturalized Suburban Garden. Photo courtesy of Gary Kennedy, 2017.

A Naturalized Suburban Garden

My partner, our two young children, and I moved to the suburban community of Pointe Claire, Quebec about two decades ago. I convinced my family of the benefits of growing first a small flower garden, and shortly after that, another even smaller herb garden in the back of the house. Having grown up in a rural setting, on a one-hundred-plus acre former farm, I was keen to have some diverse plants around the house. The two-acre expanse dedicated to mowed lawn at the upstate New York house of my later childhood had always bothered me; partly, I think, because it formed a trap for birds who would visit feeders near the house windows, becoming dangerously exposed to attacks by hawks and falcons. Or maybe it was because it just took such a long time to mow, and it always seemed to need mowing whenever there was a good hatch occurring on the trout pond behind the house. Whatever the reason, I wanted to provide our kids with some direct access to a more "natural" ecosystem.

I quickly learned that the soil was extremely poor and shallow (undoubtedly at least partly the result of decades of mowing) and that the species capable of surviving there were not especially numerous. This is not unusual for suburban developments. The normal building practice involves buying cheap farmland near the growing city, scraping off all the top soil, screening and then selling it for landscaping, and leaving the exposed subsoil as a kind of tabula rasa on which roads, buildings, and parking lots can then be imposed using a pre-theorized, cookie-cutter-like blueprint. Once the buildings are finished, grass sods or seeds are placed over a very thin layer of topsoil or sandy loam, and the regimen of ecological sterilization is then maintained as described above. We began composting upon our arrival, but there never seemed to be enough vegetable peels to make much of a difference. I began researching gardens and soils. I took bags of leaves from neighbors in the fall, diverting them from landfills to be composted in our back garden. I eventually decided to smother the grass of the entire backyard with several layers of newspaper that were then covered with about five centimeters of sandy loam. The subsoil is composed of clay and shale, and this, combined with the sandy loam and the added organic matter, has, after twenty years, resulted in a relatively good soil for growing native flowers, grasses, shrubs, vines, and trees. The sandy loam was seeded with several species of native wildflowers and grasses, although the survival rate in the early years was disappointingly low, partly because the seeds laid out on the soil surface were generally taken as an invitation to a banquet by migrating juncos and white-throated sparrows. Some species, such as pearly everlasting (Anaphalis margueritacea), did quite well in the early years and hosted large numbers of painted-lady butterfly caterpillars before eventually dying out as the soil became richer and the available sunlight receded and was replaced with the shade of growing trees and shrubs.

As the back garden continued to support increasing numbers of birds, bees, and countless other species, both above and below the soil, it became increasingly difficult and obviously nonsensical to continue mowing the front lawn. In winter, we would watch male cardinals knock the seeds of ironweeds out onto the snow for the finicky females to carefully inspect and devour. In summer, I would sit with my daughter, Martha, and watch leafcutter bees come and go from a tiny tunnel in the ground, bringing circle after circle of leaf pieces. My son, John, always loved to explore on his own, often with a stick he could use to dig. Neither these nor countless other interactions could have taken place had we not replaced the lawn with a naturalized garden. The advantages of the garden were too present in our daily experience to have us return to the insanity of the lawn mower, although I learned later that our children did experience the garden as an embarrassment in the early days. A key turning point for them was when one of their friends remarked that our garden "smelled good." The garden eventually became a source of pride for both our children.

Year after year, I have continued to plant bare-root trees and shrubs, giving the garden an increasingly wild feel to it and attracting more and more species of birds and other wildlife. I make a point of asking myself where the plant would want to be rather than asking myself where it would look right to me. There are no straight rows or consciously created palates of color in our garden. We have several Northern hackberry trees grown from seed as well as dwarf hackberry, chokecherry, service berry, winterberry, elderberry, one sugar maple, witch hazel, hazelnut, bladdernut, mountain ash, raspberries, blackberries, prickly gooseberries, and spicebush, all scattered about to seem as though they were planted by birds, wind, or other forces without any fear of God or God's enlightened agent directing them. I have also continued to sow seeds of wild grasses and flowers such as big bluestem, little bluestem, bromes, wood chesses, bottlebrush grass, June grass, beak grasses, many species of sedges and asters, bedstraw, bellflowers, dame's rocket, phlox, biennial gaura, herb Robert, coneflowers, sunflowers, milkweeds, black-eyed Susan, brown-eyed Susan, golden alexanders, beardtongues, rosinweed, cup plant, ironweed, columbine, corydalis, violets, bee balm, prairie clover, goldenrod, figworts, anemone, pasture rose, Solomon's seal, false Solomon's seal, starry Solomon's-plume, joe-pye weed, marsh marigolds, turtleheads, cardinal flower, as well as many other species that either did not survive or are rare enough in the garden that I am not consciously aware of, or immediately capable of recalling, their presence.

When our house was built, during the post-World War II economic expansion and early stages of the North American love affair with the private automobile, it was common to plant Norway maple saplings in the front yards of all the houses. Our house had a large mature Norway maple and a mature blue spruce in the front yard. The Norway maple was favored by developers presumably because of its fast rate of growth, so that a housing development could be rapidly transformed from ecological wasteland to tree-lined streets depicting order and control. The Norway maple has the added attraction of being allelopathic, meaning that its roots and leaves contain and emit chemicals that suppress the germination and growth of competing species excluding grass, acting as a natural herbicide. Norway maples also create a deeper shade than the more dappled shade characteristic of native North American species of maple. Interestingly, Norway maples are currently considered as alien invasive species, although many land developers and municipalities, including Pointe Claire, continue to offer them for residential landscaping purposes.

The Norway maple in front of our house had several large branches that were in danger of falling and had to be taken down for reasons of safety. Removal of the tree, with the end of its allelopathic toxins and the sunshine that suddenly flooded the front garden, provided new gardening possibilities. These changes also triggered eco-mental reactions in the community as the Western-Cartesian perspective was disrupted by the garden (see Figure 6.1). As the front garden transitioned from a heavily shaded garden to a tall-grass savannah, local area residents became increasingly uncomfortable. On June 15th, 2016, the City of Pointe Claire, after having received a complaint from residents with many signatures, issued a notice of infraction regarding our garden. There was a deadline of June 21, 2016 to remove the garden or face a hefty fine with escalating fines thereafter until it was brought within the guidelines of existing bylaws. The existing regulations at that time allowed for grass kept mowed to a certain height, with the option of a few well-defined and well-maintained ornamental flower beds or shrubs for decorative purposes only.

I reacted immediately to the notice of infraction by requesting a meeting with the mayor and our city councillor, neither of which were ever granted. I then started a blog about our garden and our trouble with the city, including an online petition to request changes to the existing bylaw regulations that would make naturalized gardens permissible. I expressed my intention to contest any fine that might be issued by the city in court, pointing out to the city that there had already been several similar cases decided in favor of naturalized gardens in courts across Canada. I also argued that the existing bylaw was *ultra vires* because the city's jurisdiction was limited to issues of safety and health, not aesthetics. On both counts, we felt confident that our garden was superior to mowed lawns. I continued to press for a meeting, and the deadline for the garden's eradication came and went without any fine being issued. Our family hired a lawyer who was eventually able to initiate a conversation with the people tasked with sustainable development within the Department of Public Works. These apparent allies within the municipal government managed to convince the city council to hire a biologist to do a study of the garden.

The biologist arrived at our home with an escort from public security, presumably for protection in case we became violent. We had a wonderful conversation and although the public-security agent found the conservation fascinating, she left after it became apparent to her that she was not needed. The initial conversation focused on our gardening philosophy and methods. The exercise of bringing the implicit theory behind the emergent practice of naturalized gardening and rewilding was very instructive. I was grateful for the conversation and the exercise in articulating what had only emerged at a more instinctual level. Throughout several conversations at that time, our family came to recognize that our garden had become an important factor in the construction of our individual and family identities. Our awareness was always distributed to some degree throughout the garden, either when we were in it or looking outside through the windows, to watch for birds or animals or to notice flowers in bloom. The biologist could not help but be impressed with the diversity of plant species, birds, insects, and other animals found in the garden. The copy of the report (Sardi, 2016) we received, although heavily redacted by the city with entire multi-page sections removed, clearly played an important role in convincing the city to re-evaluate their position on naturalized gardens.

The city did, in fact, eventually amend its bylaw to allow naturalized gardens such as ours. Our neighbors, on the other hand, remained divided on the issue. Some had played important roles in supporting us in our lobbying efforts. Some made signs expressing support for the garden, which they hammered into the ground where passersby could read them. We received about one thousand signatures on the online petition. Others offered kind words and encouragement to keep fighting. Without the many signs and gestures of support we received from many quarters, I do not think we would have been able to persevere. Many who we suspect signed the initial complaint to the city remain unconverted and are still clearly disappointed with the outcome to this day. The city did eventually offer an unofficial apology for what they put us through and told us that they would like to see other residents follow our example, although no official policy seems to be forthcoming that would take any concrete steps to encourage or assist residents to make similar changes.

Bodies

Mowed lawns discipline bodies, whereas naturalized gardens partner with bodies. The mowed lawn is a needy and unruly student that never seems to learn its lessons. It never stays watered, or weeded, or neatly trimmed and requires nearly continuous mowing and fertilizing. In return, mowed-lawn surfaces offer an unfettered mobility to bodies, a uniformly smooth and unobstructed plane in all directions. Mobility in naturalized gardens, on the other hand, requires negotiation (see Figure 6.2). The naturalized garden distracts one's attention with interesting things to look at, touch, listen to, or smell. In the naturalized garden, there are often small mammals hiding, possibly rabbits, woodchucks, skunks, squirrels, or cats. It is a good idea to walk quietly, but not too quietly, through a naturalized garden. I was sitting in a chair reading a book in the back garden of our house one sunny spring morning when I felt something walking over my feet. Fortunately, I did not make any sudden movement but simply looked down to see a mature skunk wobbling off to its den in the front of the house. Naturalized gardens never cease to provide surprises to bodies who learn how to cooperate with them.

Naturalized gardens without wide paths force bodies to slow down and proceed in a tentative way, checking for spider webs or patches of beggar-ticks. There are always interesting things to find, if you take the time to pay close attention. Often, I have come across plants that sprouted from seeds planted years before and had been long forgotten. The naturalized garden creates a lively form of intersubjective experience in this way—like discovering from your teenage children that they really were paying attention to the things you told them when they were five year olds. Bodies become responsive and aligned to lively landscapes in very different ways from the way the indifferent silence of the mowed lawn represses and impoverishes the senses of bodies.

Figure 6.2. Slow Down. Photo courtesy of Gary Kennedy, 2017.

A lot depends, though, on the openness of the individual to the full experience of the naturalized garden. Three-year-old Ben, for example, was initially afraid of the tall plants, but after an hour exploring the garden with Uncle Peter, Ben became more than reluctant to go back inside for lunch. On the other hand, I have been told by many area residents that if I do not already have ticks with Lyme disease in my garden, I soon will, because dogs often get ticks in the nearby nature park. For many local-area residents, our garden is a place of fear, anxiety, and confusion. Many cognitive templates are available for structuring one's relationship to a novel landscape, and normally the prevailing template is not consciously chosen but rather emerges from a historical inheritance that is edited to context when necessary to align with one's personal biographical experience and personality. Ben, like most three-year-olds, had been taught to wash his hands before eating, to pick up his toys, and so on. Places that would seem to fit into the category of "messy," combined with a profound unfamiliarity, would naturally seem dangerous or wrong upon an initial encounter. That association of danger with the unfamiliar garden is more often demonstrated as visceral discomfort than as the product of a rational calculation. This is probably even more true for adults who encounter such unfamiliar places for the first time. It is important, though, to keep in mind that places are never encountered simply by individuals. Individuals always bring a set of sociocultural mediational means with them that are put to work in the process of making sense of, and perceptually, cognitively, and emotionally reacting to, any previously unexperienced encounter with a place.

Tools

Most of the material tools required to maintain a mowed lawn are well known and easily recognizable for their associations with the lawn-mowing gestalt: lawn mowers, weed eaters, edging and trimming tools, gloves, work boots, spades, leaf blowers, soil aerators, grass seed and fertilizer spreaders, sod cutters, weeding tools for extracting the tap roots of dandelions, pesticides, fertilizers, sods, soils, and various soil amendments. A lawn maintained with goats or sheep would be fundamentally different from a lawn maintained by mechanical mowers. Each involves a very different form of material engagement.

Then there is the tool set learned to deploy the many cognitive mediators needed to actively participate in the Western-Cartesian charade of mastery that the mowed lawn represents. That learned capacity transforms otherwise normal and ordinary people into "lawn-people" (Robbins, 2007). Such an ecology of mind is organized around a specific set of sociocultural mediational means, rather like an ecosystem of plants and animals is organized around a type of bedrock, such as limestone or granite. This point has much to do with the ways in which agency is spread and distributed throughout any community of practice or any ecological system assemblage. The anxiety and economism of the Western-Cartesian paradigm make the mowed-lawn regime almost inevitable, and the relational ontology of a post-Cartesian paradigm leads one inexorably toward the naturalized garden and in the direction of sustainability.

One point that I need to make specifically about the tools used in the practice of either mowed-lawn or naturalized-garden ecologies is that the material characteristics of the entire suite of lawn-specific tools reflect, to some extent, the psychological settings that are engaged by putting such tools into practice. Anyone who has used a lawn mower knows that a certain amount of attention is required to perform the task of mowing a lawn. The task involves following the line of the already mowed part of the lawn while making sure that the mower blades do not miss any of the grass blades. But mowing a lawn is not a mentally challenging task—not as mentally challenging as, say, flaking chips off a rock to make a stone hand axe, or making a basket, or almost anything else one would make by hand. It is rather more a process of making something by machine, where the machine performs an important part of the cognitive activity. The same can be said of the various other machines that participate in the process of maintaining manicured suburban lawns: the weed-eaters, hedge trimmers, chain saws, and leaf blowers. The process involves primarily applying the implicit logic of the machine to the target, as opposed to a creative process of materially extending a mind into environment to form an active and deliberate partnership.

Naturalized gardens require far fewer material tools. What I use the most is a bucket to place bare-root plants in cold water while I plant them, or for mixing seeds with sand to then be broadcast by hand into the garden. There is a difference between the material tools used for the mowed lawn and those needed for a naturalized garden that reflects a very different form of relationship. With the possible exception of a spade or trowel used for planting, the tools of the naturalized garden are not tools characteristic of violent interactions. They do not involve the kind of implicit cognitive dissonance that a lawn mower or weed eater does, whereby the ecological destruction is in some sense absorbed by the machine. Much of the labor involved in maintaining the naturalized garden involves pulling nonnative invasive species such as European buckthorn, purple loosestrife, garlic mustard, or Norway maple. The "tool" in this case amounts to observations and the resultant mental maps of other places where these invasive species have been allowed to crowd native species out and become near monocultures rather than biodiverse habitats. A good naturalized garden requires close observation and memory in order to get a feel for how things are progressing and how the nonhuman community is reacting to the changes taking place. The naturalized garden, in other words, requires cognitive tools for active empathetic participation, whereas the mowed lawn imparts a rote or even ritualistic enactment that invites a forgetting.

Places

The mowed lawn becomes an everywhere place and at the same time a nowhere place. A mowed lawn in Montreal is supposed to look the same as a mowed lawn in the suburbs of Mississippi, Montana, Manchuria, Madinah, Monte Carlo, Montevideo, or Melbourne. Place-making within the mowed-lawn gestalt becomes an exercise in the erasure of difference and diversity. The goal is to erase the personality of a place rather like replacing names with numbers works to erase the humanity of individual people in institutionalized settings. The numbers that form street addresses perform roughly the same function in suburbs. Formerly unique and diverse ecological assemblages become parcels of real estate and abstract units located by Cartesian coordinates in economic space (Weiss, 1995). The relevant features of place become only those with economic (marketable) importance—place is determined, disciplined, forged, and fashioned by the dictates of market conditions within the unsustainability ecology of mind. It becomes almost as difficult for a suburban homeowner to have a healthy relationship with the land as it is for the guard at a prison camp to have a normal human relationship with an inmate. Clearly, neither relationship contains many of the elements that would feature in a healthy parent-child dyadic relationship.

A naturalized garden is a dynamic place where the forces of ecological succession are not actively suppressed. People and place develop and age together. This type of relationship invites a kind of rootedness that actively interrupts the omnipotence of market forces that might dictate a theoretically frictionless trading up or down of residence depending on the inhabitant's changing economic fortunes and spatial opportunities. Such diverse habitats also become much more difficult to know, compared to the mostly uniformly distributed sameness of the mowed lawn. The naturalized garden requires one to spend time to get to know when things are going well and when problems arise. Often, what appears as a problem turns out to not actually be a problem at all. The oxeye sunflowers that did so well in the front garden had always attracted many red aphids to their stems from late summer onwards, for example. Some years were worse than others, but I never did anything to suppress the aphids, thinking that eventually a ladybug, praying mantis, or beetle or some other predator would find them and the problem would take care of itself. What eventually happened was that native species of sparrows, warblers, and finches played this role during one fall migration. Although the many seeds in our garden always attract birds in the fall, that specific year the birds seemed to devour at least as many aphids as seeds and the following year the numbers of aphids were very greatly reduced.

One can speculate on how the birds experience our garden, for example, as a place of nourishment along a migratory pathway. It is clear, though, that birds do not experience mowed lawns in the same way humans do. The *umwelt* of a mowed lawn in comparison to a naturalized garden is also clearly quite different for nearly all the respective habitants. Even soil bacteria and fungi will find markedly different affordances in the two respective environments. The mowed-lawn regime itself will also offer varying chances of survival for soil microorganisms, depending, for example, on whether the grass clippings are removed or not; whether pesticides, fertilizers, or soil amendments are used or not; and so on. Some birds can reap benefits from organically maintained mowed areas. The mower often purees insects and other protein sources that then become easily accessible to foraging birds. It is also likely that just as there are organisms emerging through normal evolutionary processes capable of digesting some forms of plastics, the growing surface coverage of mowed lawns presents opportunities for some species. Skunks have learned to forage grubs, goldfinches often enjoy dandelion seeds in an open setting where predators cannot approach without being detected, and so on. These species-specific opportunities are, however, vastly outnumbered by the many foreclosures represented by the mowed lawn.

Processes

The most interesting process of the suburban landscape takes place at the interface between the ecological and psychological systems. The mowed lawn is every bit as much a part of an extended mind as a blind person's stick or a child's bobo doll. The mowed lawn provides an important part of the curriculum for the implicit learning and identity construction of the contemporary ecological psychopath. It provides perceptual, cognitive, and emotional tools characteristic of an orientation toward confrontation, whereas the naturalized garden redirects that orientation toward a partnership model. The mowed lawn, like anthropogenic plastics, epitomizes a compulsion to mastery—the normalization of imposing one's will upon an unwilling Other. The mowed lawn acts as a symbol of the Western-Cartesian historical reflexive experience of trauma, anxiety, shame, and fear and the socioculturally organized, ritualistic response in the form of aggressive regimens of suppression and control.

The process itself is unavoidable, but these dysfunctional Western-Cartesian characteristics of the process may be open to remediation. The naturalized garden represents, of course, only one facet of a much more complex ecology of eco-social-psychological unsustainability. The already-lived-in world of our births is the world we interact with to carry out our psychological development. It is the world that educates us continually. That already-lived-in world is replete with the reflexive makings of anxiety and fear and trauma-driven human action. The process of ecological-psychological correspondence calls us to respond to the implicit anxiety and trauma already in the world as we find it. In responding, we continue the ongoing cycle. To upset the process, we would need to respond not to the world as we find it but rather as we might imagine, with our best selves, what we and the world might become reflexively together, in a parent-child type of partnership.

Agency

A mowed lawn and a naturalized garden operate as two very different types of agents that act upon peoples and on the world. The lawn or naturalized garden forms relations with the property manager/gardener that are in many ways like a parent-child type of relationship. Obviously, the human acts upon the landscape, whether by planting seeds, applying fertilizer, compost, pesticides, water, irrigation systems, pink flamingos, or garden gnomes, for example, or by removing dandelions and grass clippings or any other routine human-land interactions these places entail. But, the lawn or garden also molds the human, establishes those same corporeal routines, directs attention, and provides cognitive tools and specific forms of communication flowing in specific directions. The lawn or naturalized garden is, at the same time, like a parent and like a child.

Aside from the occasional worm-hunting robin or the grub-hunting starling, mowed lawns tend to repel birds, offering little more than a place to rest temporarily. Seeds and berries are absent by design. Lawns are antisocial agents that enforce only formulaic routines, minimizing any chance of incidental learning and existing solely for the mythological currency it provides. The capacity of the lawn to enforce such mindless routine on suburban peoples depends partly on a destabilizing ecological reality, where the monoculture becomes increasingly difficult to perpetuate, and partly on the governmentality of suburban cultural norms—in other words, a myth.

That myth includes the notion that the person mowing, watering, and eradicating the species diversity of the mowed lawn is in charge and is controlling the fearful wilderness, keeping the potential for chaos at bay. In fact, if we look closer at the suburbanite lawn mowers, what we find is the opposite: the lawn disciplines the people, reducing the potential for cognitive connectivity with a living world. The peoples' way of being in the mowed-lawn world incorporates a routine of ecological destruction while the person is pressed by the aesthetic social norm to repress the consequences of their own action. The agency exercised by the lawn meshes particularly well with the market economic system as it extends the illusion of individual consumer sovereignty.

Although the belief in consumer sovereignty persists in capitalist societies, consumers do not really exercise very much agency in the capitalist economic system either. Each individual consumer looks to the Other for guidance and instruction. Robbins (2007) illustrates this point well with the example of the pesticides and fertilizers needed for the maintenance of the suburban lawn and the pattern that also maintains the economic system:

> Industry, which does indeed depend on my purchase of these chemicals, is not inventing this desire, at least not in some way separate from the immediate visceral experiences of daily life. It is rather reflecting and enforcing the images formed in the experience of lawn ownership. Community pressures, which are so essential to the continued use of inputs, are also not a simple product of the demands of industry. Rather, it seems apparent that individual and community desires respond most directly to *the demands of turfgrass*. Neighbours respond to the needs of Poacea, not to shareholders, as they stare over their back fences at our brown patches and dandelions. The lawn makes these demands. (p. 133; italics in original)

This point brings us back again to the hallucinatory nature of capitalism, where agency appears to reside in rational actors (people), but in fact can more realistically be found in the objects of our desire and the cultural scripts setting the parameters of acceptable (capitalist) behaviors. Although embedded in a rhetoric of freedom, the capitalist subject surrenders agency to the imagined dictates of grasses, automobiles, advertisements, and fashion (always according to the imagined-as-inevitable dictates of the market). But perhaps most of all, agency is surrendered to and exercised by fear and anxiety. Fear of the wild and untamed contagion—the uncontrolled—and the neurotic compulsion to spread instructions, control, and safety drive capitalism. In the end, it is the promise of a mechanical world, operating according to knowable laws and the dogma of mainstream economic theory that the capitalist subject willingly if unwittingly delivers agency to.

Memory

The mowed lawn represents an attempt to control nature by freezing it in a timeless and placeless abstraction of imaginary safety, one where ecological processes are ritualistically neutralized in the hope of forcing them to cease to operate without close guidance and control. The mowed lawn, like so many other materializations of the Western-Cartesian ideology, represents an attempt to repress the memory of past trauma, to remake the world into one conceptualization or another of a timeless, placeless human-designed safe zone. From the destruction of forests as a ritualistic enactment of the myth of bringing the subconscious under control, to systematic destruction of "uneconomic" wetlands, to the design of suburbs as geometric grids oblivious to ecological specificity, the pattern of Western-Cartesian person-land relations is one that is above all a reflexive trace of past and continuing traumatic confrontations.

The mowed lawn loses its memory, loses its voice, and ultimately is restrained and prevented from errant movement of any kind. This is also the reflexive impact the mowed lawn turns back on its captors. Homeowners are disciplined by the regime required for this ecological petrification. With their attention focused on the routine—the recipe and not the pot—the value of diverse and psychologically stimulating places is forgotten, connections are broken, and synapses atrophy. The suburbanite becomes increasingly powerless to recover traces of sustainability. Fry explains:

> "What is decided" is that in coming to recognize the omnipotence of the unsustainable there is an essential and total change in the frame of reference of ethics that reframes the question of "what we are and do." Rather than diminishing the significance of human conduct, such a change makes it all the more important. So rather than "making of the past what you will" the creation of a "future with a future" filters how "the past" is viewed, because it can help expose what futures and what defutures. In other words, the postmodern axiom of undecidabilty is stood on its head. (Fry, p. 109)

All lawns have pasts, presently erased but still potentially accessible as pasts and futures. The replacement of mowed lawns with more ecologically valuable practices will be, in many ways, like the truth and reconciliation process currently underway in Canada. The residential-school system emerged from a dysfunctional framing of Other as lesser (Plumwood, 2002); in the same way, the mowed lawn emerged from a dysfunctional framing of land as lesser. The

Figure 6.3. An Alibi. Photo courtesy of Gary Kennedy, 2017.

pathway to sustainability, then, involves bringing that way of framing, that mediational means, out into the light for examination and re-evaluation in order to put the memory back into the land and to recover the story only a healthy landscape can tell.

Narrative

Whether the suburban landscape manager is cast in the role of hero or villain depends largely on the conceptual tools available to the community, but it is important to recognize that these roles are also determined to some extent by the roles played by supporting actors. When the land is cast in the role of mowed lawn, its character cannot be anything but confrontational. The lawn can only play the role of the villain because normal ecological processes can do nothing but work against the efforts and aesthetic norms of the lawn mower. There is, in other words, a real world operating on its own terms whether our stories want to admit as much or not. The dominant Western-Cartesian narratives have been woven over many hundreds of years on top of and in relation to that real world. That means essentially that fail-safes are built into the narratives to guard against incidental ecological learning. The pattern of action that drives collapse—collapse, followed by trauma, followed by action that doubles down to lead to a bigger collapse—is maintained largely through narrative.

Anomalies of incidental learning must be dealt with for the system of Cartesian dualism to be maintained. Suppression of learning is greatly facilitated by the length of a normal human lifespan and by shifting baselines as we naturally view the world through the prejudicing prism of human time, but in recent decades, the rate of change and the scale of collapse have become so omnipresent as to strain the cultural blind spots, making the situation more difficult to suppress. The emerging tribalism in Western politics, the increasingly fanatical obsession with seemingly transcendent technologies such as the colonization of space, the fantasy of Kurzweillian immortality (Kurzweil, 1999), the retreat into screen cultures and virtual realities, and the rising incidence of social distress are all signs of the real world pressing up against the limits of human psychological capacities.

As the toolmaking, tool-using, and tool-constitued species, our evolutionary fitness is tied to our tools, including the narratives we use to hold the prevailing culture together as a cohesive system. The mowed lawn, too, creates a tension between the culturally enforced version of the way the world works and the disappointingly dismal outcomes that result from being guided by such a blatantly inadequate and Sisyphean story. The Western social imaginary, however, is well practiced in the mantra of "there is no alternative" even when it comes to the transformation of a not-insignificant percentage of the planet into that ecological wasteland known as the lawn.

When sanity does break out, the Western-Cartesian ecology of mind works to keep it ritualistically contained. Ritual, defined as the collectively practiced enactment of the dominant mythic-cultural narratives, acts as a sort of default setting, removing the action from any serious questioning. It is just the way the world works. Only in so far as different and more challenging approaches, like the naturalized suburban garden, are allowed, would we need to find ways of fitting them into the dominant perspective as exceptions that prove the rule. We construct alibis for them. These sorts of alibis (see Figure 6.3) work incessantly to cover up traces of sustainability that continue to simply pop up as the real world continues its one-sided conversation with itself.

The naturalized garden cannot simply be "black-boxed" and ignored because the mowed lawn carries such a powerful symbolic message. The tentative trial-and-error process of finding a satisfactory explanation usually begins with the hypothesis of some defect in the homeowner. Laziness, stupidity, physical incapacity, and mental illness are all likely candidates for solving the naturalized-garden puzzle. These are also, not coincidentally, the historic templates generally employed when encountering non-dualistic, non-Cartesian cultures. The idea that a garden might have constitutive value or that the garden could reflect itself into my identity is so completely antithetical to the Western way of being in the world that such an explanation simply remains beyond comprehension. The land itself does not make a good candidate for explaining the anomalous landscape feature because from within the Western-Cartesian paradigm, the land is supposed to receive instructions and not the other way around. Residents are pushed to invent comprehensible stories that can mesh with their already-mastered narratives about the way the world works. The usual solution is to turn to the conceptual tools of consumer sovereignty and the indifference curves of consumer preference.

In the case of our naturalized garden, a friend of our family paid the National Wildlife Federation to have our garden certified as "wildlife habitat" and delivered the sign to us so that we could frame the garden for our neighbors within a narrative that would be comprehensible to them. Unfortunately, by putting the otherwise sustainable artifacts into narratives so that they become accessible via unsustainability tools, the real story gets lost. That story must be told with all the tools of a relational ontology to inform the creation and maintenance of the naturalized garden. By framing the garden as "wildlife habitat," it is brought back into the master-servant gestalt of unsustainability. The garden becomes comprehensible as a provider of psychic income to homeowners exercising their consumer sovereignty. Once reassuringly explained and cognitively situated within the unsustainability complex, the garden can be comfortably ignored, its power to disrupt reassuringly neutralized.

Atmosphere

Familiarity and deep cultural resonance provide mowed lawns with a patina of pride and contentment when the rules are followed and ecological forces are temporarily overcome. The mowed lawn is like the loveless parent who drives the child to redouble efforts to gain parental approval and affection. Unfortunately, the child's efforts often only make the child appear more pathetic to the bad parent. The tension of the double-bind relationship between either child and parent or homeowner and lawn forms the unrecognized background atmosphere that may go completely unnoticed, may occasionally subside, but never entirely goes away.

The atmosphere of the naturalized garden, on the other hand, becomes a conversation, an explicit zigzag of give and take, of making stories together. The naturalized garden is full of action; there is always something interesting happening for those who look. In winter, rabbits transform the garden into a midnight restaurant where items on the menu include prickly gooseberry, spicebush, hackberry, hemlock, blueberry, raspberry, and—when those are exhausted—chokecherry and bladdernut. Birds visit feeders and flower and grass heads in their search for seeds in winter. When spring arrives, hatching and hibernating bees emerge and crawl out of the ground. An uncountable variety of insects shows up. Crows begin gathering large sticks for nests, followed by many other resident birds that content themselves with less ambitious nesting materials. Transient spring-migrating birds stop for something to eat and a place to rest before moving on. All summer long, the garden undergoes a continuous process of transformation as flowers, fruits, and seeds develop and mature, ensuring the evolving continuation of the garden (see Figure 6.4). When fall arrives, seeds are self-sown, insect eggs are laid, leaves fall, and mushrooms come out. Throughout the oscillating calendar,

Figure 6.4. A Refugia. Photo courtesy of Gary Kennedy, 2017.

memories are refreshed. Hello, catbird. Hello, towhee. Hello blue cohosh and mycelium. A recovering ecology renews old friendships, and the atmosphere begins to contain much more than can be spoken in words. Rhythms of color and movement overlap and interplay continuously. The prolificacy of births in spring is sometimes punctuated with the staccato thud of a juvenile catbird striking a pane of glass or the slow staggering dance of the squirrel afflicted with raccoon roundworm. At other times, the melodic unfolding of a single leaf of the tulip tree can sing the opening lines of the entire human story.

References

Baumgartner, M. P. (1988). *The moral order of a suburb*. New York, NY: Oxford University Press.

Dale, A., & Newman, L. (2010). Social capital: a necessary and sufficient condition for sustainable community development? *Community Development Journal*, 45(1), 5–21.

Fry, T. (2015). Design in the Maelstrom of time. In T. Fry, C. Dilnot, & S. C. Stewart (Eds.), *Design and the question of history* (pp. 97–110). New York, NY: Bloomsbury Academic.

Guthman, J. (2003). Fast food/organic food: Reflexive tastes and the making of "yuppie chow." *Social & Cultural Geography*, 4(10), 45–58.

Jenkins, V. S. (1994). *The lawn: A history of an American obsession*. Washington, DC: Smithsonian.

Kurzweil, R. (1999). *The age of spiritual machines: When computers excess human intelligence*. New York, NY: Penguin Books.

Larsen, L., & Harlan, S. L. (2006). Desert dreamscapes: Residential landscape preference and behavior. *Landscape and Urban Planning*, 78(1), 85–100.

Plumwood, V. (2002). *Environmental culture: The ecological crisis of reason*. London, UK: Routledge.

Robbins, P. (2007). *Lawn people: How grasses, weeds, and chemicals make us who we are*. Philadelphia, PA: Temple University Press.

Sardi, M. (2016). *Rapport de l'évaluation écologique du terrain situé au 166, avenue Sunderland, Pointe Claire*. Pointe Claire, QC: Unpublished municipal report.

Simmel, G. (2000). *Simmel on culture: Selected writings*. London, UK: SAGE.

Tokarski, M. (2019). *Hermeneutics of human-animal relations in the wake of rewilding: The ethical guide to ecological discomforts*. Cham, Switzerland: Springer.

Weiss, A. S. (1995). *Mirrors of infinity: The French formal garden and 17th century metaphysics*. New York, NY: Princeton Architectural Press.

· 7 ·

TRACING THE ABSENCES OF SUSTAINABILITY

> My search has led me to a novel idea of human knowledge from which a harmonious view of thought and existence, rooted in the universe, seems to emerge. (Polanyi, 1966, p. 4)

> That is the sort of world we live in—a world of circuit structures—and love can survive only if wisdom (*i.e.*, a sense or recognition of the fact of circuitry) has an effective voice. (Bateson in Forge, 1973, 249)

We, within the Western-Cartesian paradigm, live in a world without sustainability, without the brains, bodies, cultural/cognitive tools, material culture, or even environments needed to guide us out of our zone of proximal development (ZPD) and to imagine, support, institutionalize, and maintain sustainability minds. To achieve sustainability, these traces must be recovered from the many absences of sustainability we experience daily and then rewoven into a more sustainable way of becoming with the world. It has been helpful for me, throughout my research and studies aimed at understanding the challenge of sustainability, to consider another very personal absence from my own life experience as a metaphor for the many absences that ultimately constitute the contemporary unsustainability.

On April 25, 1960, my mother, Martha, died at the age of 24 of acute pericarditis, creating an absence in my life that continues to this day. I was

22 months and 20 days old at the time. An absence, though, is not simply a void. An absence is something more than nothing. It leaves traces, and I have tried to get to know my mother over the years by following those various traces. The process of coming to *know* my late mother through her traces has formed my template of self-discovery and the ongoing pattern of my life. When I was eight or nine years old, standing in line at a supermarket with my stepmother, a woman I had never seen before was compelled to come up to me to tell me what a wonderful person my mother was, how everyone loved my mother so much and could depend on her for good advice and friendship, and on and on until I was so uncomfortable for my stepmother that I could no longer process the words that the strange woman was saying to me. I had several experiences like that one throughout my childhood, but not usually in the presence of my stepmother. I continue to be grateful for each one of those people who tried to help me with clues and traces to fill in my absence. My stepmother obviously experienced the absence of my birth mother very differently from my own experience, and our respective experiences of absence were directed through very different pathways. My father would often turn to alcohol to deal with the grief of his own experience of absence, especially around the anniversaries of Martha's death. On some of those occasions, he would explain to us that my younger brother, David (about seven months old at the time) had killed our mother by making her get out of bed to change his diaper. Other times, he would explain how his own mother had murdered Martha with a comment about her own ability to care for her three children by herself through good health and bad. (We had been staying at my grandparents' house at the time of my mother's death, while our house was being painted.)

I remember finding a school report my mom had prepared for one of her high-school classes. I don't know whatever happened to it. I was expelled from the family shortly after my father's death, and whatever artifacts existed from my mother were not passed to me. The report had a wooden cover she had made by hand. It was bound, but I don't remember how. The pages were typed on a typewriter, with a few corrections made by hand. The report was on dogs. There were hand-drawn illustrations of several different breeds of dogs. They were quite beautiful and very skillfully drawn. It was clear to me that my mother had loved dogs, and the report reflected a very mindful and meticulous caring. My grandmothers and grandfathers told me stories about my mother from time to time. My father's parents told the most stories. I think it was often too painful for my mother's parents to remember and to reflect on the daughter they had lost. Martha's parent's stories were often about a little girl, sometimes riding a bike, picking flowers, or playing with friends. My father's parents told stories about a generous, strong, confident, and well-loved young woman who loved babies, people, and the outdoors.

I eventually attended college in Massachusetts within easy hitchhiking distance of my mother's sister, my aunt Dorothy. I went to visit her over a long weekend in October. She took me to visit several museums in the Berkshire Mountains and made plenty of time for conversation as we rode in the car from one museum to the next. I had not seen my aunt for many years, and the experience of being with her was uncanny. I noticed several subtle behavioral gestures, patterns of speech, and other more ephemeral signals that gave me the impression of looking in a mirror. I had the distinct feeling of being closer to my kin on that long weekend than I had been throughout my remembered childhood. She explained to me at one point that she had assumed the care of my two brothers and I immediately following my mother's death, but that my father had ended that arrangement on the grounds of her refusal to include spanking as part of a toilet-training regimen. These various experiences of seeking out the traces of my mother have had a formative effect on my thinking around the issue of sustainability. I cannot help but draw parallels between my formative experiences and a society whose formation and development is guided by a now configured-as-hostile world. Traces of sustainability are often removed from our awareness by people in authority who act according to one cultural script or another, no matter their intentions or understanding of their own actions. Just as the healthy development of a child is impinged by spanking, the sustainable development of society is impinged by a commoditized and anxiety-based world.

Sometimes the land would tell me a story about my mother. Here is one of those stories. One early summer day when I was eight or nine years old, my brothers and I were tasked with picking some wild strawberries for my stepmother. The local wild strawberries in our area grow to be just slightly smaller than a pea. They did not grow abundantly anywhere on the dominantly clay soils of my father's property, ensuring that filling the strawberry containers would take several hours to accomplish. Through trial and error over several strawberry seasons, we gradually came to recognize the likely places to find strawberries. When all the known strawberry patches had been visited on that occasion without making much of a dent in the containers we were supposed to fill, we headed over a hill toward a small creek at the end of a pasture where some adolescent hemlocks grew among pin cherry, poplar, green ash, and yellow birch to search for snakes, salamanders, and crayfish. Near the top of the hill, at the edge of our property and next to the neighbor's plantation of red pine, we noticed a slight clearing in the meadow and could see British soldiers, blue-eyed grass, and hawkweed through the sparse meadow grasses. We decided to check it for strawberries on the way to the creek.

My younger brother saw them first, but we all soon began to find them through the entire area of mosses, hawkweeds, and sparse, wispy grasses. We had stumbled upon a large patch of cultivated strawberries, like treasure hunters finding an unexpected chest of gold. We quickly filled our baskets while also eating many luscious berries. We forgot about the creek and headed home like conquering heroes. When our stepmother initially saw the baskets full of large, cultivated strawberries, she was suspicious. Her initial thought was that we had visited a neighbor's garden, a few miles away, and stolen the strawberries. As was her habit when it appeared we had accomplished some dastardly deed, she sought out our father to witness, confer with, and assign an appropriate punishment. Our father listened to our story of discovery, and I noticed that a sudden change come over his face when we told him where we had found the strawberries. It was as if a ghost had entered the room. He explained that Martha had planted strawberries in that area shortly after they purchased the property. We never spoke further about those strawberries, and we were never asked to pick strawberries again.

This chance encounter with a trace of my mother was quite informative. The strawberries were not near the house but rather they had been planted quite a long walk away, through what had been a cow pasture or hay field at the time. The strawberries were planted in a place where they could thrive on their own within a pre-existing ecology. This trace of my mother provided a clue to a mind that put the world into a particularly connective and respectful way of being. That order held the promise of long walks with children through meadows to collect strawberries, weaving and binding a family to a place of wonder and abundance. That ordered way of being could continue to "speak" to my brothers and I even after my mother was gone. It spoke about the importance of good knowledge, the importance of putting strawberries where they would be comfortable and able to use their own voice. It told me that my mother thought it was important for her family to have a treat of fresh strawberries in the spring.

The social mind is very much about the presences and the traces of absences being etched into minds made of brains, bodies, tools, and places. To come to know sustainability, our best bet is to focus on the traces of what is missing within the Cartesian and neoliberalized social mind, those absences that can only be encountered indirectly. What, for example, could the experience of the pre- or post-marketing mind, the pre- or post-enclosures, or a world of shared commons have been like, or be like in future? What could this imagined experience tell us about sustainability? Ecologically speaking, those pre-invasion, Turtle Island communities can never be recreated. Too many immigrant species, including earthworms, supply-demand models, luxury sedans, nanoparticles, plastics, and video lottery terminals have now closed that door forever. Some absences may have been missing for so long now that from the current perspective they might as well have never existed at all, yet they might make themselves known today or in future in such forms as an undefinable collective heartache, a longing for a butterfly, a bird, a smell, or a feeling no longer directly accessible. Unsustainability has, in a sense, left us all parentless and orphaned, without any person or thing to take our hands and lead us out of our own unsustainability ZPD. We might productively seek out whatever traces exist in a hoped-for future without falling into the trap of a utopian linear development, of isolatable causes and effects. Such a future would focus not on the strawberries, but rather the relationships formed that brought forth the strawberries.

I often gain insights into my mother through her namesake, my daughter. "Dad," she asked as a six-year-old, holding her father's hand on the walk to school, "Are we here for a reason?" The question is not as important as the relationship that allowed it to form. Watching my daughter grow into a strong, thoughtful, caring, and intelligent young woman has often given me insight into my own mother. The likeness between my mother's marriage portrait, given to me by my grandmother, and my daughter's high-school-graduation portrait is uncanny. My daughter's athleticism and choice of sports—hockey, lacrosse, and rugby—her large network of friends, and her generous and caring nature, are all consistent with the other traces I have found and nurtured over the years of coming to "know" my mother. Watching my daughter's childhood often brought to my mind the stories I had heard from my mother's parents about her childhood.

Sustainability is now as absent and dead to the world as my mother has been to me, with only traces left in the form of a sort of vague promise. If we wish to come to know, feel, and become sustainability person-tool-environments, we must all together look to those traces of what once may have been, what could have been, and what might yet become. This process of painstakingly searching for traces represents a deliberate choice of becoming and remembering as well as a deliberate choice of forgetting. The process is constitutive and has clear identity implications. It involves selecting some tools and thoughtfully foregoing others. It will be a process of self-discovery based more on inner reflection than on unchallenged belief in the conventional wisdom.

Within the family where I spent my childhood, there was a regime of truth established about who I was. Continually and obsessively trying simply to fight that false narrative instituted a personal process whereby I began to enact and become an angry and psychologically damaged person. The struggle to be sustainable within an ideational and material context of unsustainability can be exhausting. The kindest thing I think my stepmother ever did for me was to expel me from the family following my father's death. I was able then to begin a process of self-reflection and healing. There will, however, be no such escape from the economistic mind or the techno-anxiety of the contemporary world, so the task of becoming a sustainability society can be expected to be much more difficult. We are all like fish unknowingly swimming in the waters of unsustainability.

I do not think the solution lies simply in the adoption of a traditional-knowledge way of being in the world, either. The lessons from Mezirow's transformative learning theory (see Chapter 2, pp. 67–68) are important in that respect. It will be much more important to challenge and undermine the conventional wisdom, forcing a remaking and rebuilding of collective and individual identities from scratch than it will be to find a different way of being in the world that could theoretically be adopted and worn like a new set of clothes. Settler identities absolutely need to confront their pasts and incorporate that past into their revised image of self. This is not primarily about shaming but rather about moving from hallucinatory persons-lost-in-spaces to becoming people who take responsibility for the constitution of a partnership/sustainability mind. The past cannot be escaped in any case once it has wormed its way into the mediational means, forming and re-forming persons-in-environment. In as much as the narratives we tell ourselves about who we are form and constitute us, it is important to recognize that there will never be a new story of the future without a new story of the past. This will involve confronting the same past with new tools for understanding. Sustainability will continue to remain foreclosed without attention to the past. Fry, Dilnot, and Stewart (2015) build on this point about redesigning the past rather more eloquently than I ever could:

> The challenge before us thinkers of, or about design, is to hold the ground of the advancement of "history" counter to consciousness. What this means is holding the space of possibility opened by Heidegger's recognition of the interpretation of history centering on the historical being an ontologically foundational structure. The nature of this structure is not recoverable in itself, rather it arrives in the form of interpretations emanating from a new ontology, which itself is an effect of the practice of transformation. "History" (the historical) here is placed before design as agency—as historical events that in significant part ontologically design subjects that interpret them (the past). Historical agency is therefore lodged in foundational events that, in part, bring them, as subjects, into being. It follows that there is simply not "us" and historical events *for we are historical*—even though "we" (in all our differences) have little or no awareness of our being as such. As historical beings we are a trace of more than the past experiences of our "self." (p. 13)

I cannot escape my childhood any more than the Western world can escape its history of colonialism. My own personal identity has undergone a radical transformation, but it did not occur without also a radical transformation of not only the constitutive story of my past but also an ontological transformation of myself and my material/historical context. The person I am has been reconstituted and redesigned from the same historical events through a shift to a relational ontology. Understanding myself and my mind as mediated, extended, situated, embedded, enacted, distributed, and dynamical incidentally rewrites my historical past (Malafouris, 2013, p. 58). As design theorists, Fry, Dilnot and Stewart normally focus on that material context, but in this passage the focus floats across the sociocultural matrix from the narrative aspect of history and the taken-for-granted aspect "history" plays in the (un-) sustainability identity, to the ontological transformation that redesigns the historical to enable that redesigned historical to then, in turn, redesign the present and future.

Malafouris (2013) makes another important point in this regard: "materially enacted metaphors, unlike their linguistic expressions, present no text-like propositionality. They do not simply communicate meaning; they communicate actively doing something" (p. 65). The metaphor that environment is like a parent, once enacted in material culture, becomes much more broadly communicative than we generally recognize from within the Cartesian-tooled historical inheritance. Just as mowed lawns draw us into a world of unsustainability, a world of trees and gardens materialized as partners, using their own voices and recognized as partners, would also draw us into a world of sustainability.

Traces of sustainability can be uncovered by examining historical events with a new relational ontology and the psychological and material reconfiguration those events triggered. The simple act of making that connection with a new paradigm-shifted ontology already leads us in the direction of sustainability. Getting to a point where one is willing and able to turn their ontological gaze inward also represents, for me, a hugely transformative first step. The works of educational, social, and developmental psychology theorists such as Bateson (2002), Bronfenbrenner (1979), Bruner (1986), Lave and Wenger (1991), Rogoff (2011), Vygotsky (1978, 2012), Wertsch (1991, 1998), and Wertsch, Del Rio, and Alvarez (1995) have been enormously helpful to me, not only in theorizing (un-) sustainability but in providing guidance to put myself, including especially my historical self, under the microscope to work out the dimensions of my own ZPD in order to seek out good guidance and to work at healing myself. The examination of historical events is conducted not simply to find out how the world works to improve our management and control of it; it is a fundamental exercise for understanding who we are and how we were acquired by the specific identity of the unsustainable. We *can* be, and we *would* be, as different as the worlds we create and that, in turn, create us simply by focusing our attention on some traces while ignoring those that are harmful. If we want to stop the insanity of unsustainability, we need to stop driving the world insane so that the world, as we have configured it, can, in its own turn, stop driving us insane. Sustainability is always about mandala and ecology of mind, whether the loops are running through the historical or the emotional, where the shape of the pathway is cyclical and not linear.

There is a pain implicit in some absences. The loss of a parent is not always necessarily detrimental, but it is almost always painful. The pain emerges mainly from the idea, the cognitive tool, one uses to think about and practice an idealized image of the parent-child dyad that is foreclosed by death. My life was dominated, before my father's death, by that idealized image of recovered relationship with my father that was never to come about. The foreclosure has as much to do with the image of self as it does with the image of parent. Sustainability is a type of absence that can best be truly known intuitively, implicitly, through these sorts of visceral traces. Forcing a rationalized, language-based understanding only works to foreclose a more complete and realizable encounter with sustainability. I have no memory of my mother, yet my entire being continues to ache for knowing her. Part of my knowledge of my mother is revealed to me through that visceral absence. The pain honors the image of my foreclosed mother-child dyadic relationship. To honor the earth entails a similar type of pain and honoring.

Yet some pain can be so chronic as to simply form the constant background noise of one's quotidian life. Most people I interact with daily are not aware of the absence I experience, not aware of the traces I am constantly seeking out, whether traces of my mother or the traces of sustainability. Many are not consciously aware of their own pain inflicted by the absence of sustainability or other absences we all experience. Certainly, there is some level of conscious awareness of absence across society and an understanding at some level that things are not going as well as the economists would have us believe. Many who are unable to confront that explicitly and subconsciously employ defensive mechanisms to police what information they attend to. There is, however, a widespread, measureable anxiety and distress around the issue of sustainability, particularly in the over-developed world. Many people have told me they prefer not to know the ecological state of the world. Of course, that unwillingness to confront reality, ironically, makes our species' obsession with brain size somewhat moot. It undermines even the effectiveness of the Western scientific, positivist project. There is obviously no guarantee humanity will survive into the twenty-second century, but if we, as a species, are incapable of making a concerted effort to seek out the traces of sustainability, then one must ask whether conscious awareness has any evolutionary-survival benefit whatsoever. There is pain in the world, but pain can be as constructive as it can be destructive. The Western-Cartesian way of dealing with pain has historically been to seek ways to erase it. Sustainability will entail ways of reconfiguring pain into assemblages that honor the many processes entailed in the ontological shift to a societal-scale partnership mind.

It is also difficult for this process to get very far if we undertake it only as individuals. The partnership mind, environment-as-parent mind, needs to also be a social mind. Much of what I know about my mother has been gleaned from conversations with many people who did know her. Traces of sustainability can often be found by weighing outcomes of past decisions, uncovering the traces of peoples who did know sustainability, challenging the myth of progress, and reframing ourselves as peoples who are "willing to learn" (Bateson, 2010, p. 88). The alternative is to continue an impoverished existence, on autopilot, deferring all decisions in our lives to the markets, lawnmowers, highways, plastic straws, and everything else we tacitly accept as our masters. If I had been well nurtured by perfect parents, would I have still devoted so much of myself to the search for my mother's traces or the search for understanding of the dilemma of sustainability? Maybe not, but the question is purely hypothetical in any case—nothing will change the fact of my mother's absence. The only variable left is how I choose to deal with that absence. It is also hypothetical with respect to the search for the traces of sustainability. What kind of persons do we truly want to become? What kind of world and what kinds of tools would we need to become those person-tool-environments? When we can answer those questions truthfully and courageously, then we might be able to begin recovering traces of sustainability and institutionalizing sustainability across all our mediational means, and all the parts of the situated, extended, and distributed social mind.

Traces can sometimes be found by considering intentions. I have learned much about my mother by thinking about the choices she made within specific contexts, but also about the choices her presence and absence have induced in others. The absence of sustainability does change the world, but it is still the world, and the zigzag process of the world continues just the same. We normally guard against anthropomorphizing the world from within Western academic traditions, but from a biosemiotic perspective, we can say that in our world processes do tend to unfold in a dialectical or zigzag pattern within a given directionality. The presence or absence of sustainability does not change that background or higher scale direction. A life is also lived within a similar higher scale pattern of intention. That background intention, whether a person or a world, can be characterized by the term "becoming with." I opened this book with a quotation from Gregory Bateson (Figure P.1), explaining what the term "ecology of mind" meant. We cannot keep all ideas and all knowledges alive; we must choose which are worthy of our becoming with the world and which ones should wither on the vine. It is worth remembering that the world we inherit is pregnant with intentions. Plastic and mowed lawns deliver a false sense of safety and security by pretending to "master" brute nature. Trees intend something else, and we can sometimes witness the evolution of those intentions as the character of a tree and its community change over time. Our encounters with hummingbirds, leopard frogs, spring rains, and sunsets intend to transport us and transform us into a different world. We are often not consciously aware of our intentions, but when we configure the world to restrict its potential connectivity to mowed lawns and maybe a few cockroaches and spiders in our homes, then our intentions become obvious all the same.

I have had many experiences, especially throughout my childhood, of a sense of my mother's presence coming to me through the land. Caring for the land comes, in some sense, very honestly and very naturally for me. I grew up in a family that was so obviously dysfunctional that it became not just easy but also imperative for me to believe that "many worlds are possible" (Bruner, 1986, p. 149). At times, I suppose my existence depended on that hope, on the implied promise of that sentiment. A visit to France at the age of 15 with my high-school French class also confirmed that there existed other ways of being in the world. And knowing from a very young age that my mother was literally located in the soil no doubt provided me with atypical cognitive tools or metaphors to frame my relationship with the land. I would not, however, recommend my own specific pathway to acquiring a model of subjectivity consistent with sustainability. Although the lessons from my experience are clear, they are not as readily transferable through what we, in the Western world, would consider to be education. A process of abstract knowledge transmission cannot be trusted to induce the required transformative changes. It will take guidance, to be sure, to move us as a society beyond our current ZPD toward sustainability, but that guidance has always been accessible to us. What has been missing, in part, is a willingness to ask for guidance, to stumble upon the strawberries, figworts, or brook trout and to listen to their story. The anxiety stems, I think, from an always-looming specter that we might slide backwards to the untamed and uncivilized side of the conceptual universe, to need instruction and guidance, and to become weedy and not useful. The fear must be overcome one way or another. Fear is a primary anchor holding unsustainability in place.

More than seeking good guidance though, we need to pay close attention to what we take *as* guidance. On this point, Ingold (2013) makes a useful distinction between anthropology and ethnography (pp. 2–4). Ethnography, according to Ingold, involves "the work of documentation," whereas anthropology is described as something more akin to coming to know the marinade through a process of self-maceration. Documented data or propositional knowledges will not take us very far in the direction of sustainability. Sustainability will always involve being in a world that reconstitutes us as sustainability-tool-people-world assemblages. Traces, presenting, for example, as strawberries or clown-faced blow-up plastic bobo dolls, need to be skillfully put into a sustainability practice. The current problematic *absence* of sustainability likewise needs to be practiced with that entire world in mind. To my father, absence was practiced as a witch hunt, seeking out the *cause* of the absence, whether soiled diaper or mother-in-law's sharp rebuke, and then assigning blame, with little or no thought to the collateral damage being left in the wake of such an inquisition. Parallels can be drawn, or perhaps my father's practice was simply a microscale example, a kind of ritualistic enactment, of the practice of Western-Cartesian paradigm. The best advice I can offer on this front comes again from Ingold (2013):

> Truly, never in the history of the world has so much information been married to so little wisdom. To me, wisdom runs not in straight lines but along the ways of the donkey. This humble beast, slow but nimble, cross-eyed but big-eared, capable of traversing the most rugged ground without mishap, without need of road or rail, or fossil fuel, has served humankind for many thousand years. Whereas the automobile, once it has reached the end of the road, can go no further, the donkey just keeps going. Non-human it may be, but should we not pay heed to what it has to tell us? All true scholars are donkeys: obdurate, capricious, dogged, curious, petulant, at once captivated and astonished by the world in which they find themselves. They will not be hurried but go at their own pace. They live in hope, not in the illusion of certainty. Their paths may go this way or that, unpredictably. They find the grain of things and follow it, and in so doing, find themselves. All learning, as I hope you have found by now, is self-discovery. Where next? *Know for yourself!* (p. 141)

Personally, I would interpret Ingold's advice to "know for yourself" as a call to understand the contingent nature of the mediational means you have been immersed in, and mastered by, and to always remain open to the kinds of paradigm shift that would leave those tools by the side of the road and replace them.

Nothing in the preceding pages and chapters will *necessarily* bring us any closer to sustainability. All the words above are open to attack from the "epistemology police" (Latour, 2004, p. 253) and may, under such scrutiny, be used also as a tool to blindly reinforce a dogma of unsustainability. It will be important to "know for yourself" and to make that donkey-like practice of knowing a way of honoring the self-world dyad. As Bateson (1991) warned at the outset in his description of an ecology of mind, "ideas are interdependent, interacting … ideas live and die. The ideas that die do so because they don't fit with the others" (p. 265). So many alternative knowledges have been dismissed out of hand throughout the entire history of Western civilization, because of the double-binding blind faith in the one, and only one, true knowledge. Of course, any knowledge, whether tacit and embodied or cognitive and rational, propositional or narrative, will conjure one world into being while simultaneously erasing other potential worlds. So many worlds have now been foreclosed by the drunk-driving Western epistemology that now, at this moment in history when our one-dimensional world is about to collapse, we stand collectively on the precipice grasping almost blindly for traces of an ethos of sustainability. They will not be found where we have been looking with our Western-Cartesian minds.

Some absences leave traces, like strawberries or hummingbirds hidden in plain sight, appearing only when we are ready and open enough to find them. If we think about the stories we tell ourselves about mowed lawns, or the economy, or plastic, or nanoparticles, or democracy, it is perfectly obvious not only that all these stories are missing something but also that the alternative narratives are immediately accessible and ready to be put into practice, should we ever so choose. One of the more dangerous aspects of the Western-Cartesian historical heritage is the acquired adeptness at conceptual compartmentalization. To some extent all cultures bequeath "epistemologies of ignorance" (Malewski & Jaramillo, 2011), but the Cartesian method of ignorance generation and dissemination seems to be particularly pernicious. The Western-Cartesian perspective includes many learned incapacities. We learn to be incapable of imagining life without flush toilets or iPhones, while simultaneously becoming incapable of considering these inventions in their proper *context*. A proper context would include the whole historical journey, including mistakes, traumas, misunderstandings of mistakes, misunderstandings of traumas, new mistakes, new traumas, and so on until finally an impoverished world impoverishes itself further with toilets and iPhones.

At our home in Pointe Claire, there are no strawberries in our naturalized garden. My children had to make do with blueberries, blackberries, and raspberries … and poison ivy, prickly gooseberries, and beggar-ticks. The traces I intended for them did not include strawberries. As children, John and Martha would always get excited when hummingbirds would come around at the fishing camp in the Laurentians. I planted native columbine, corydalis, cardinal flower, bee balm, phlox, and other species known to attract hummingbirds. We put up hummingbird feeders, but whole seasons would slip by without a single sighting. We almost always had one species or another of hummingbird moth hanging around our garden, but very rarely would we spot an actual hummingbird. I had read that hummingbirds often migrated with sapsuckers to steal the sap from holes drilled in trees before other food sources became available in the spring. One year, after sighting a sapsucker drilling holes in the neighbor's Chinese elm, I hung a hummingbird feeder at a time when crocuses and snowdrops were in bloom. There was little else a hummingbird could get nourishment from, and it seemed a little bit crazy to think a hummingbird would be anywhere within a thousand kilometers. But, within two or three minutes, a hummingbird appeared at the feeder and stayed at the feeder for about five minutes, before leaving to steal sap or follow the sapsucker north, never to return. There were odd sightings in odd years, on bleeding hearts, jewelweed, cardinal flowers, or turtleheads, but hummingbirds began to seem about as real as unicorns or fairies after years of fruitless waiting and watching. It was only when I planted many inconspicuous figworts, after both children had grown up, lost their childish fascination with hummingbirds, and moved away, that Shirley and I began to see hummingbirds on a regular basis. I did not even plant the figworts for the hummingbirds. I had read an article suggesting that part of the problem with declining bee health and numbers was a loss of diversity in their diets.

That's how, I think, sustainability might one day just show up, albeit at a societal scale, as a person-tool-environment response to an essentially empathetic gesture without a quid pro quo, without being forced, and without a direct and instrumental intention, like hummingbirds simply responding as hummingbirds do, to recognizable traces in a place that just seems finally all right. We have only to think of how we currently react as a society to some of the obscene absences that currently confront us, (the lost Atlantic cod off the coast of Newfoundland, grayling in Lake Ontario, phthalate-free babies, iPhone-free teenagers, or missing wolves in Pennsylvania, etc.) to realize how daunting the sustainability challenge we face really is. The first step, however, may simply be to recognize our infinite absences for what they are, to see those absences in the world as reflections of our own absences. Then, may we find the courage to extend our field of concern and empathy, not through a process of planning or remediation, not as part of a bargain, but rather as a simple gesture of honoring the world so that, one day, just maybe, the world may also once again honor us.

Bodies

The absence of sustainability gets lodged in bodies in some obvious ways: malnutrition in the deserts of the Sahel, food deserts in Chicago, or obesity, diabetes, or other diseases associated with over-processed, industrial-model food. It is somewhat less obvious in traces of DDT, phthalates, and other uninvited toxins coursing through the blood, organs, and muscle tissues of every living thing that swims, crawls, or walks across the face of the earth. Even less obvious are the mental-health impacts that the absence of sustainability has on our bodies. According to Gifford and Gifford (2016), "Climate change can affect mental health even before its actual appearance. Distress related to impending environmental change, such as habitual ecological worrying and 'eco-anxiety,' has been increasingly noted" (p. 292). The world, in other words, currently does not provide good enough parenting for sustainability bodies to "crystalize out" from the "interstices between objects" (Gosden, 2005, p. 197). Bodies carry the stresses and anxieties of their ecological contexts even when the stress is not consciously and collectively acknowledged.

I remember the moment distinctly, sitting in the Stanley Tavern in Montreal, when I realized I felt healthy for the first time in my life. I was an American undergraduate studying economics at McGill University in 1988. I could pay for the final year of my undergraduate studies with savings earned in the States from jobs in residential construction and in restaurants. At the time, I wondered what it could have been about the food at the Stanley Tavern, but in hindsight, I realize it was probably the first time in my life I felt safe. I was beyond the reach of my abusive father. I had escaped the predictable tide of criticisms and innuendo from my stepmother about my many perceived flaws and inadequacies. I see no reason why the ontological shift to a sustainability mind would not bring a similar sense of security and health to the social mind and to the collective bodies suffering unknowingly from the ongoing unsustainability. Certainly, if we consider only the social distress caused by the collapse of biodiversity or by global climate change, for example, and imagined what it would feel like if, for even one day, our politicians, journalists, and teachers dealt with the issue as an absence, then I think we could imagine an entirely different world.

Tools

We live in a Western-Cartesian world with an absence of sustainability tools. We have neither the stories of sustainability, nor the architectures, nor the artifacts, nor the vocabularies, nor the conceptual models, nor the models of subjectivity consistent with sustainability. In addition to our absence of sustainability tools, the tools we do put into practice are complicit in ongoing processes of the creation and maintenance of sustainability absences. Increasingly sophisticated electronic video games and virtual-reality experiences respond directly to perceived absences. Increasingly, anxiety and mental illnesses are routinely addressed pharmaceutically. The pharmaceutical approach turns the focus away from underlying causes and secures the double bind. In Europe and Asia, the practice of "forest bathing" has been enlisted as a tool for addressing the absence of sustainability (Beilinis, Takayama, Boiko, Omelan, & Bielinis, 2018; Hansen, Jones, & Tocchini, 2017). These tools and many others respond to the absence of sustainability but do not work to uncover the traces of sustainability. That is an important distinction that needs to be kept in mind.

Forest bathing might, within the psychology of the marketing mind, result in more urban nature parks and more people having direct access to places that appear to be uncontrolled by humans. Of course, nothing could be farther from the truth. Nature is simply conceptually enlisted in a different form, as commoditized therapeutic resource, currently being maximized as a therapy for improving the productivity of labor, but contingently available for the construction of shopping malls and high-rise apartments when market conditions change. Forests for forest bathing are economistic coping tools and not tools for uncovering traces of sustainability. The forest loses its voice as a forest within the constellation of forest bathing; all the people can hear is the instrumental value of a doctor-like or pharmaceutical-like healer. To allow the forest to simply "be" a forest would violate the deeply subconscious Western-Cartesian obsession with control and mastery.

Process

The process of uncovering traces might, from a different perspective, be simply referred to as *learning*. But learning to repair the damage requires insight into the nature of the damage and the ability not to fall prey to "a temptation to confuse the idea of manipulation with the idea of a cure" (Bateson (1991, p. 268). This is a critically important point about the process of learning for a cultural/ecological transformation. Once it becomes codified in "best practices," or as "thou shalt" prescriptions for naturalized gardens or for traditional ecological knowledge boiled down to a drinkable or injectable essence, it simply falls apart and "goes dead."

> The relationship I have with my mother, for example, is not materialized in photograph or headstone, it is not restricted to an anniversary. It is, in some sense at least, a *living* relationship, with all the openness and contingency that term implies. That type of relationship cannot be imposed as a model from above. It must emerge, grow, and develop organically, being cared for and nurtured along the way, while at

the same time other potential forms of relationship are left in abeyance, dis-activated through neglect. This is not to say that the job of birthing a new sustainability into shared practice is impossible—only that it is impossible when the approach is formulaic. Jeffries (2011), for example, provides good and relevant advice to grief counselors:

> One human soul sitting with another who is suffering the pain of grief. A care provider who understands that the grief journey is different for each individual. A care provider who is sensitive to different cultural and religious/spiritual backgrounds, and understands that your function is to support the journey and not control it. (p. 2)

"Supporting the journey" might well involve some knowledge of the process of reconstituting constellations when an important part of that constellation transforms from presence to absence, but the "human soul … who is suffering the pain of grief" cannot be forced into a prescribed new constellation, no matter how comforting its implicit promised future may be, without it eventually falling apart. The process of recovering traces of sustainability will always be a contingent and unfinished process, in direct confrontation with the Western-Cartesian obsession with closing off by disproving hypotheses, by nailing things down, by splitting the world into categories of either true or not-true. As much as we may have been mastered by deeply subconscious fears and anxieties about the zigzag nature of all the processes of our becoming, the fact remains that the more we fight and rail against what we misperceive with our inadequate cognitive tools to be frightful, and the more we strive to gain the upper hand, to take control, and to deliver the world into bondage and slavery, the more the world simply falls apart, becoming more lethal even as we continue proclaiming our victories.

Agency

Recovering traces of sustainability requires recognizing distributed agency and the importance of tacit understandings. Plastics, private automobiles, televisions, and other screen technologies all work to induce the delusion that the world is infinitely malleable and pliably controllable. Like mowed lawns, they break healthy connective, cognitive, and emotional bonds to a community of life and extinguish habitats and partnerships of becoming. If these processes had been consciously designed, they would fit the criteria of Tony Fry describes as "defuturing" (Fry, 2012). But consciously designed or not, there is little doubt about the magnitude of their influence over the speed and direction in which human civilization is currently evolving. These objects resist the multitude of efforts currently being directed at uncovering a way of being in the world that allows human continuation in a form worth enduring. They enlist us from birth as child soldiers in the war on nature.

The trick, then, might be to consciously work to stop supporting the ubiquitous agencies of the dynamo, while simultaneously supporting instead the agencies of strawberries, figworts, or birds who plant mountain-ash trees or poison ivy. Why not form partnerships with birds and other species, inducing wonder in their simple yet profound ephemerality. This shifting toward a more parent-child like relationship with environment might then be nurtured and be extended into the fields of architecture and design by asking: How can this building, this place, or this thing become a better parent to me? How can it make me a better sustainability person?

Memory

Memory, engaged for the purposes of uncovering traces of sustainability, is more productively thought of as a *practice* and not a collection of data or knowledge(s). It entails a deliberatively targeted way of remembering. It recognizes the constitutive agency of the processes of uncovering, remembering, and honoring. It entails remembering the absences, the what-could-have-been, but is instead foreclosed by plastic, private automobiles, coal mines, personal screen devices, industrial-fishing trawlers, video lottery terminals, cement sidewalks, or throw-away fashions. Memory for sustainability becomes a process of reimagining and renegotiating, of listening instead of talking, and of coming to new agreement; it begins from a deep understanding that many worlds are possible and the materialization of one specific world arises with remembering.

Memories will also need to be coaxed into performing parenting duties if our goal is sustainability. For example, headstones might be replaced with trees, ecological damage with restored habitats. Bodies might remember how to walk with the world rather than on it. Workers—whether high-paid executives, janitors, or bulldozer operators—might remember how to avoid doing things they would not want their children to see. Students and teachers, planners and architects, and trade-deal negotiators and Wall Street bankers, all need to relearn how to remember the context they are operating within and remember their roles in the enactment of the world.

Narrative

Stories consistent with recovering traces of sustainability are fundamentally about listening attentively to, rather than imposing one's voice and one's will onto, a lively world. It is fundamentally a story about giving a voice back to our many partners and relations—all the things and beings that cooperate in our co-becoming. It is also about learning to tell the difference between the voice of a hummingbird and the voice of a video lottery terminal. It is the story from the first lines of this book, a grander and greater story whose first lines might have been written by Vico (1744/1968):

> For that first language, spoken by the theological poets, was not a language in accord with the nature of the things it dealt with (as must have been the sacred language invented by Adam, to whom God had granted divine onomathesia, the giving of names to things according to the nature of each), but was a fantastic speech making use of physical substances endowed with life and most of them imagined to be divine. (pp. 127–128)

Vico was not afraid to face a world with tools that can provide only partial and mediated poetic forms of knowing. The story Vico tells is a story of learning to live in the middle, between the zig and the zag.

Bateson and Bateson (1987) add the necessary storylines to Vico's tale to form a workable narrative for uncovering the traces of sustainability:

> It is part of human nature to learn not only details but also deep unconscious philosophies—to become that which we pretend—to take the shape and character our culture imposes upon us. The myths in which our lives are embedded acquire credibility as they become part of us. Such myths become unquestionable and are built deep into character, often below conscious awareness, so that they are essentially religious, matters of faith.
>
> It is to these myths—and the future form that they may take—that all our mythmakers, our mythopoets, including scientists and politicians and teachers, owe responsibility. The doctors and lawyers and the media share responsibility in the dynamic myth—the answers they offer to the Riddle of the Sphinx. (p. 182)

The Riddle of the Sphinx, "What is a man that he may know a number, and what is a number that a man may know it?" (Bateson & Bateson, 1987, p. 177), is where any story must begin. It establishes the pre-ethical foundation for any narrative by carving the world into comprehensible pieces. The Western-Cartesian perspective answers the riddle in a way that cuts peoples off from their whole selves, establishing the double-bound regime of unsustainability.

The first step to sustainability must be to rewrite the story of who we are, and that new story must redefine the human to include the mediational means, things, and objects that ultimately come to determine who we can become.

Atmosphere

Absences can be faced with intentional forgetting, using any number of coping strategies from minor distractions, to complete immersion in a life's work on the one hand, to extremely destructive addictive behaviors or psychopathological conditions on the other. They can also be faced head on, but the social costs of the absence of sustainability simply cannot be avoided, as Gifford and Gifford (2016) remind us:

> Eco-anxiety is characterized by severe and debilitating worry about risks that may be insignificant and is not associated with the more proactive behavior associated with habitual ecological worrying. It can elicit dramatic reactions, such as loss of appetite, sleeplessness, and panic attacks, among those affected. Even without these symptoms, many people are often unconsciously anxious, unaware of just how concerned they truly are about impending environmental changes. This lack of awareness often slowly diminishes, to be replaced by increasingly severe anxiety about how dire the situation has become. The National Wildlife Federation has estimated that 200 million Americans will eventually experience emotional distress as a result of the effects of climate change. (p. 292)

Western society has been internalizing these types of costs from the start—they are our story. There is no baseline for comparison. The underlying orientation of fear and absence takes a toll, but that price is impossible to quantify.

What will it feel like to look at this era in human history if the average predictions of biodiversity collapse by the end of the twenty-first century are borne out? What stories will be told to children when the oceans have been emptied of fish and the only "seafood" left to eat is jellyfish? Leaf blowers, lawn mowers, and weed eaters may not have yet come to dominate the spring soundscape when Carson (1962/2002) worried about a "spring without voices" (p. 2), but when children only look for birdsong on Google, and the word *atmosphere* invokes fear of climate change rather than breath and air, we can be quite certain that sustainability is dead.

References

Bateson, G. (1973). Style, Grace, and Information in Primitive Art. In A. Forge (Ed.), *Primitive art & society* (pp. 235–255). London, UK and New York, NY: The Wenner-Gren Foundation for Anthropological Research.

Bateson, G. (1991). *Sacred unity: Further steps to an ecology of mind*. New York, NY: Harper Collins.

Bateson, G. ([1973] 2000). *Steps to an ecology of mind*. Chicago, IL: University of Chicago Press. (Original work published 1972).

Bateson, G. (2002). *Mind and nature: A necessary unity*. Cresskill, NJ: Hampton Press.

Bateson, G., & Bateson, M. C. (1987). *Angels fear: Toward an epistemology of the sacred*. New York, NY: Macmillan.

Bateson, M. C. (2010). *Willing to learn: Passages of personal discovery*. Hanover, NH: Steerforth.

Bielinis, E., Takayama, M., Boiko, S., Omelan, A., & Bielinis, L. (2018). The effect of winter forest bathing on psychological relaxation of young Polish adults. *Urban Forestry & Urban Greening, 29*, 276–283.

Bronfenbrenner, U. (1979). *The ecology of human development: Experiments by nature and design*. Cambridge, MA: Harvard University Press.

Bruner, J. (1986). *Actual minds, possible worlds*. Cambridge, MA: Harvard University Press.

Carson, R. (2002). *Silent spring*. Boston, MA: Houghton Mifflin. (Original work published 1967).

Fry, T. (2012). *Becoming human by design*. London, UK: Berg.

Fry, T., Dilnot, C., & Stewart, S. C. (2015). *Design and the question of history*. New York, NY: Bloomsbury Academic.

Gifford, E., &Gifford, R. (2016). The largely unacknowledged impact of climate change on mental health. *Bulletin of the Atomic Scientists, 72*, 292–297.

Gosden, C. (1994). *Social being and time*. Oxford, UK: Blackwell.

Gosden, C. (2005). What do objects want? *Journal of Archaeological Method and Theory, 12*, 193–210.

Hansen, M. M., Jones, R., & Tocchini, K. (2017). Shinrin-yoku (forest bathing) and nature therapy: A state-of-the-art review. *International Journal of Environmental Research and Public Health, 14*, 851.

Ingold, T. (2013). *Making: Anthropology, archaeology, art and architecture*. London, UK: Routledge.

Jeffries, J. S. (2011). *Helping grieving people—When tears are not enough: A handbook for care providers* (2nd ed.). New York, NY: Routledge.

Latour, B. (2004). *Politics of nature: How to bring the sciences into democracy*. Cambridge, MA: Harvard University Press.

Lave, J., & Wenger, E. (1991). *Situated learning: Legitimate peripheral participation*. Cambridge, UK: Cambridge University Press.

Malafouris, L. (2013). *How things shape the mind: A theory of material engagement*. Cambridge, MA: MIT Press.

Malewski, E., & Jaramillo, N. (2011). *Epistemologies of ignorance in education*. Charlotte, NC: Information Age.

Polanyi, K. (1968). *Primitive, archaic, and modern economies*. Garden City, NY: Anchor Books.

Polanyi, M. (1966). *The tacit dimension*. New York, NY: Anchor Books.

Rogoff, B. (2011). *Developing destinies: A Mayan midwife and town*. New York, NY: Oxford University Press.

Vico, G. (1968). *The new science of Giambattista Vico*. Ithaca, NY: Cornell University Press. (Original work published 1744).

Wertsch, J. V. (1991). *Voices of the mind: A sociocultural approach to mediated action*. Cambridge, MA: Harvard University Press.

Vygotsky, L. V. (1978). *Mind in society: The development of higher psychological processes*. Cambridge, MA: Harvard University Press.

Vygotsky, L. V., (2012). *Thought and Language; Revised and expanded edition*. Cambridge, MA & London, UK: MIT Press.

Wertsch, J. V. (1998). *Mind as action*. New York, NY: Oxford University Press.

Wertsch, J. V., Del Rio, P., & Alvarez, A. (Eds.). (1995). *Sociocultural studies of mind*. Cambridge, UK: Cambridge University Press.

· 8 ·

CONCLUSION—TRACING AHEAD

I began this book by arguing that to achieve sustainability, we first need to understand environment as our parent and then to establish an understanding and practice of the pre-ethical foundations of our common historical inheritance of mediational means, which are currently cultivating, guiding, and disciplining the unsustainability ecology of mind. It is worth pointing out that we have become essentially orphans in our relationship with environment, regardless of what our Cartesian inheritance would have us believe. The current pre-ethical foundation, or "bulk of life," (Gosden, 1994, p. 125) does not need to be accepted as a historically determined and unalterable given—"many worlds are possible" (Bruner, 1986, p. 149).

Although it has nearly always formed the cultural model of the assumed-to-be-objective way the world works, it is now clear that there is never a completely objective perspective from everywhere. Any knowledge is always a situated knowledge and any mind is always a situated mind. The Western-Cartesian worldview and the relational ontology offered by Material Engagement Theory (MET) provide two very distinct perceptions and understandings of ourselves and our world, including our collective past, present, and future. Although the current globalizing worldview is built upon pretensions of access to objectivity, such accessibility to the world, even with the wonders of Western science, is simply not in the cards for the human or for any other Earthly species. Our perception, understanding, and feelings about the world are always derived from the way our minds are parented and educated by the already-lived-in world of our births, our own *umwelt*. Haraway's (2016) words bear repeating:

> It matters what matter we use to think other matters with; it matters what stories we tell to tell other stories with; it matters what knots knot knots; what thoughts think thoughts; what descriptions describe descriptions; what ties tie ties. It matters what stories make worlds, what worlds make stories. (p. 12)

We *can* learn to uncover traces of (un-) sustainability and use those traces to make other tools sets, other matter, other knots and ties, and other stories and worlds of sustainability. The first step of that process involves shifting our ontology away from the mind-and-matter splitting Western-Cartesian paradigm. If we can achieve that shift on a large enough scale, it will begin to seem insane to continue to destroy our own extended and distributed selves as our modern ritualistic practices currently prescribe. The solipsistic economistic and terrorized mind will change, and the underlying anxiety will dissipate. By continuing to pay close attention to the characteristics of our relationships with the land, with other species, with knowledges, and design practices—in short, with all the various elements that constitute our eco-mental triadic systems—sustainability can become thinkable, sensible, story-able, danceable, draw-able, sing-able, expressible through all the forms of design that design us—all the masterable forms of action that in turn master us.

The difference between Western societies and sustainable societies is not primarily, as Diamond argues, a difference in ecological affordances in the forms of plant and animal species amenable to being brought under human control. Rather, it is the underlying learned and delusional compulsion to master and take control that separates the Western-Cartesian unsustainability from the potential sustainability mind. That difference includes a linear, "billiard-ball gestalt" (Clark, 2002, p. 6) understanding of the way the world works so that Western people fail to notice that mastery goes both ways. The Western perspective misperceives and misunderstands the very nature of relationship and good parenting. The inevitable geographical sorting, segregation, and concentration that was part of the package deal of domesticating plants and animals opened Western society up to a pattern of collective and recurring psychological trauma. Trauma became and continues to be a dominant aspect of the Western ecology of mind, as diseases and pests ravage monocultures of crops and domesticated herds of animals—especially when those diseases jump from animal to human to cause abnormal outbreaks of mortality and suffering. It becomes apparent in arms races and rape cultures, deforestation and toxic by-products, landfills, plastic toys, and universities. The pattern has worked its way into our economic conventional wisdoms and other mediational means to the point that the systems itself *depends on* continuous trauma and disturbance. This pattern, of course, continues to this day, as it is sunk so deeply in the Western psyche as to have become its defining characteristic, including all the extended instances of that mind distributed throughout environment.

There are certainly aspects of the Western mind that might still be worthy of respect and preservation, if not reverence, but clearly, taken as a unitary system, the Western mind is in dire need of a radical overhaul. Mowed lawns would make a good place to start the spring-cleaning, but radical changes are needed in our approaches to education, regional and urban planning, design, research and knowledge production, and the many materializations arising with these processes. The pathway to sustainability will, I think, need to get much more personal and introspective. It will need to press home the tortuous absences that unsustainability always implies and then materializes. Our approach to parenting and education needs to become much more like the approach of Indigenous peoples. For example, Swentzell (1997) describes this relational approach as follows:

> This world, for me as a child, was very comfortable and secure because it gave a sense of containment. We roamed in the fields and nearby hills. At an early age we learned an intimacy with the natural environment and other living creatures. We learned of their connectedness to rocks, plants, and other animals through physical interaction and verbal communication. We gained tremendous confidence and an unquestioning sense of belonging within the natural ordering of the cosmos. Learning happened easily. It was about living. In fact, the word for learning in Tewa is *haa-pu-weh*, which translates as "to have breath." To breathe or to be alive is to learn. (p. 57)

This passage is deceptively complex. Swentzell is comparing the materializations of a typical Tewa pueblo with those of a day school built by the Bureau of Indian Affairs. Learning "in the fields and nearby hills" is compared, in other words, to learning in a place that "had a barbed-wire fence around its periphery. That fence defined the complex and effectively kept the two worlds separate" (Swentzell, 1997, p. 62). These tacit knowledges represent the most important traces of (un-) sustainability.

It will become increasingly difficult to avoid the perilous positive feedback loop triggered by anthropomorphizing the world into the enemy. Breaking the pattern, though, is within our grasp if we can come to understand the nature of the challenge. The survival imperative of the twenty-first century involves shifting that culture-based material engagement away from one of confrontation and toward a shared practice of parent-child-like partnership. Although new behavioral models will naturally emerge or simply resonate more from some sectors of society than from others, we need look to those peoples, tools, and places that actively engage us in turning our attention toward a world in need of help, support, listening to, and partnership. This positive outcome is only possible with a shift toward a relational ontology, a recognition that our fate and our mind is inextricably entangled and co-constituted by and with a world we are in turn co-constituting.

This recognition of being with a world is the same mindful dynamic of sustainability that often comes into play when a touchstone of identity, whether individual or group identity, is challenged with harm or destruction. Perhaps a creek or woods, iconic of a local sense of place, is suddenly threatened by economic development, and this triggers an implicit awareness that one's mind will no longer be extended and distributed as before. Activists bubble up from time to time in this way, providing possibilities for broader social learning when guidance is available to connect the dots. Society's storytellers—the politicians, scientists, journalists, and teachers—have a sacred responsibility in such cases, whether they recognize and accept that responsibility or not. This closing chapter briefly explores some of the ways the previous chapters might come to inform sustainability research and practice moving forward, as we begin the task of connecting those dots into a broader post-Cartesian paradigm shift.

Economics

As I and others have argued, economic theory begins from a set of underlying assumptions borne out of anxiety and Cartesian dualism. It paradoxically emphasizes submissive prostration to the quasi-divine power of market control while simultaneously proclaiming individual consumer sovereignty. Economics disciplines *bodies* through the conceptual transformation of the human aspect of people-tool-environments into units of salable labor at the beck and call of market forces whether as rent-collecting capitalists, salaried white-collar workers, hourly rent-a-bodies, or homeless shelter "clients." Bodies are expected to perform for the market, including retooling, re-educating, re-inventing, and relocating according to current market conditions. The standard supply-demand models used as conceptual *tools* to put into practice the establishment and maintenance of economic institutions of unsustainability exercise considerable *agency* within and throughout the *processes* of the formal and informal economic system of global-market capitalism. The implicit *narrative* of the modern market-economic system is perhaps where insanity reaches its highest level of concentration, although psychopathy, like *agency*, is distributed widely throughout the system and spills across the various domains of extended and distributed minds. What is especially lacking from the tool kit of the mainstream academic discipline of economics is any recognition or acknowledgment of economic system as both constitutor and product of marketing *atmospheres*—that is, the peculiar yet normalized way capitalism feels to us, the way the ontological fiction of control and commodity gets distributed across the various domains of the extended mind.

Fixing mainstream economics is simply a fool's errand. Sustainability requires a new approach. The '"Toxic Textbook' sticker" (Barry, 2012) should be taken very seriously:

> WARNING!: This economics textbook is not suitable for use on real planets that face ecological limits. Application of the ideas, implicit values and analytical tools found in this textbook, without high levels of skepticism and caution, can result in overheated atmosphere, greater difficulty surviving and widespread misery. Other species may experience extinction. Conflicts with commonly accepted human values. *Suitable for use in imaginary worlds only for historical study*. (p. 134; italics in original)

The basic neoclassical relationship between theory and practice cannot be rehabilitated. However, there are alternative frameworks that could represent a step in the right (sustainable) direction. Community economic development (CED) provides one example. According to Shragge (2003),

> CED is a strategy that has been developed primarily but not exclusively in response to deterioration of local economies and the lack of hope for revitalization from the outside, either from the private market through investment or with the support of government programs. (p. 123)

CED could be shifted to a relational-ontology framework to form a materially engaged and fully mindful approach to economic development. Although that step has not yet been taken, a close reading of contemporary CED literature

(Fasenfest, 1993; Greer & Gonzalez, 2017; Mendell, 1983; Shragge & Toye, 2006) could provide an excellent entry point to other interesting literatures classified as heterodox economics (Barry, 2012; Boulding, 1978; Gowdy, 1994; Latouche, 2009; Schumacher, 1974). Certain institutional economists, such as Hodgson (2001) and Polanyi (1968, 1977), have also begun to establish a theoretical framework that could be based on a relational ontology. On the other hand, dominant strands of environmental economics and ecological economics have smuggled so much of the Cartesian psychopathy into their models and theoretical frameworks that they should be only approached with the greatest of caution.

The primary challenge to the academic discipline of economics is to account for the way stories, in the forms of models, myths, conventional wisdoms, conceptual tools, material tools, artifacts, and so on, are used within the practice of economics to prevent the telling of other conflicting and potential sustainability stories. Cost-benefit analysis (CBA) is a case in point. What constitutes a cost or a benefit is determined exclusively by convention within this theoretical framework. There is generally no genuine effort to get at the pre-ethical bedrock to which an ecologically rational cost-benefit framework *might* be grounded and anchored. What is more, since only quantifiable monetary units of analysis can be considered, the dark magic of CBA engages in a ritualistic practice of converting the world into a place no one in his or her right mind would want to inhabit. CBA demonstrates clearly that economics is the mediational means for the projection of suffering onto the world (Keating, 2018).

There has probably never been a human community that did not form some commonly shared conceptualization of "economy," given that the term denotes the cumulative material consequences of intentional action. I am not proposing that economics be abolished, only that it be radically transformed to account for the matter that really matters (Gosden, 2005, p. 197; Haraway, 2016 p. 12). Organized efforts to reimagine a discipline of sustainability economics also need to include a recursive or zigzag perspective. From this new perspective, it is clear to me that economics and education can no longer pretend to operate as separate endeavors.

Education

The conceptualization of education as a distinct endeavor involving the mastery of the self emerged, in part, from the context of Western anxiety and the obsession with control. The impact of education is, in other words, indexical of the traumatic historical experience constitutive of Western civilization. It represents and guides the enactments and materializations of the Western-Cartesian mythology and dysfunctional mediational means. From some non-Western perspectives, life and learning are nearly synonymous. Education is indexical of the prevailing patterns of human life (see Nelson, 2017, pp. 49–60). Indigenous approaches to community economic development often allocate much attention of the recovery of Indigenous languages and cultural traditions. The implicit point being that Indigenous economic institutions and practices emerge organically from Indigenous patterns of thought, structured within non-Western language and cultural structures and practices. The historically emergent ontological difference between Western-Cartesian and non-Western approaches to education is well illustrated by Nelson (2017) and worth quoting at length:

> Implied within the concept of symbiosis—living together in community—there is an important emphasis on dialogue, exchange, and council. These exchanges are intergenerational, intertribal, intercultural, and often interspecies. Unlike normal classroom discussions, which can be very productive in terms of outcomes, this kind of communication usually is not based on achieving certain results (although it can be). Many Indigenous learning circles focus on a different type of learning and sharing. First, there is generally not a single, specific teacher, but all in the circle are considered teachers and learners. Second, everyone sits in a circle, making an even playing field (no one is higher or lower or privileged as special). Third, there is often no time limit on what one shares. It can be very short or very long, and there is no cross-talk or interruption, unless the speaker asks for it. Fourth, there is an emphasis on deep listening, where personal assumptions and biases are internally questioned and suspended so as to truly hear another's voice and meaning. Long spaces of silence are welcomed and considered healthy.
>
> This Indigenous learning space, often called a learning lodge, invites deeper questions that generally are not asked and engenders a collective inquiry—a type of "group mind" or group mediation through words—that is palpably different from other learning spaces. In this circle, there is shared wonder and sometimes grief or frustration at the complexity or contradiction of certain questions or topics. Most importantly, judgement and competitive ego activity are strongly discouraged in this circle, whereas deepening the capacity to hold divergent views as equal to one's own is encouraged, thus making a safe and, many would say, ceremonial or even

> sacred learning space where one can share without fear of being wrong, reprimanded, scolded, or ignored. It is shameless education. (pp. 54–55)

Furthermore, Nelson mentions the importance of materiality and the importance of establishing correct relations with that materiality within much of Indigenous learning practices:

> Learning from the land is a primary method of gaining knowledge. Many Indigenous worldviews see nature as the ultimate teacher. It is essential to learn from and be in reciprocal relations with the more-than-human world: the local plants, animals and elements that give us life, the sun, moon, wind, fire, soil. (Nelson, 2017, p. 54)

Our historical inheritance of mediational means is reflected in the way we conceptualize, materialize, and put into practice education, and that practice, in turn, reflects the ways in which the economic system is established within institutions and conventional wisdoms. It therefore makes little sense to separate and maintain two distinct academic disciplines of economics and education. The structure and character of an economy ultimately emerges from the persons-tools-environment-matrix. If people see and understand themselves as ultimately constituted reflexively within that matrix (the job of education) then a quite different economy would naturally be expected to emerge from that altered matrix. When we get serious about reuniting mind and matter in a post-Cartesian relational ontology, the combining of economic and educational knowledges will make obvious good sense.

Education, as we currently conceptualize it, would become obsolete with the shared recognition that for something to be valuable, it needs to have constitutive value. The value of education would no longer be associated with credentials used to command premium prices in a labor market. It would no longer be aimed at the development of new weapons of mass destruction for our war on nature. It would no longer be about finding theories and methods for keeping the systems that Cartesian-directed collective human action drives to fall apart to just keep going a little bit longer. Education would be about tools and environment, about recognizing good tools and learning how to use them, about seeing the best person you could ever become together with environment, and about learning to respect and accept that relationship.

I am struck by the way normal Western educational processes impose themselves on *bodies*. Primary-school children, who normally would otherwise learn in action and through play, are routinely constrained in chairs, at desks, looking forward, and having the marketing mind implanted in their bodies. The *tools* of education include, obviously, the furniture, the buildings, the playgrounds, the bell, the clock, the computers, blackboard and chalk, and the books (often treated as sacred texts to be learned sufficiently to be spit back verbatim in examinations). There is also a whole religion full of conceptual tools used to knit positivism, economism, anxiety, and education together into a whole cloth. Those tools contain and supply the answers to questions such as: What is knowledge? or What is the purpose of life? Putting those tools into practice constitutes a *process* of cultural reproduction, not unlike the practice of flaking rocks to make hand axes. Education should not be about fostering and protecting, but about disrupting current processes of impoverishment, replacing them with a more holistic process of becoming within a child-parent structure of relationship.

The *narratives* of education, with some notable exceptions, are generally the same as the stories told by economists. The theme is always about maximizing the capacity for control, whether in the form of purchasing power or in terms of self regulative capacity. These narratives form important nodes of distributed *agency*. Together this ecology of the Western-Cartesian mind forms the current *atmosphere* of unsustainability. There are many babies in the bathwater of our current educational system, but education is full of illusion, like the rest of everyday life within the Western-Cartesian disability.

Reforestation

Where ancient forests can still be recovered, they should be. Forests probably make the best parenting environments. Instituting a massive reforestation campaign is as much about the psychological health of people as it is about fighting global climate change, air pollution, or recovering sources of potable water. This is another example where the environment-parent metaphor is particularly useful. The practice of "forest bathing" (see chapter 5 above pp. 242–243) requires forests, but it forms the kind of relationship with the forest that is like the child seeing the parent as simply a bundle of goods and services, culminating ultimately in an inheritance of instrumental value rather than constitutive value. Forest bathing, in other words, represents a somewhat impoverished form of relationship with the forest, at least if we take it at face value. Certainly, it is difficult, if not impossible, for a healthy child to spend time with a healthy parent without deeper bonds forming. This undoubtedly occurs also with the practice of forest bathing, but the point

I want to make here is that the mental image is not entirely consistent with a sustainability ecology of mind.

Food systems also need to be adapted to the program of reforestation accordingly, by moving away from an industrial large-scale, long-distance, corporate-agriculture model, to a community-scale agroforestry model. I know from personal experience that this change is worth fighting for. The shift from ecologically-impoverished-isotropic-plains-peoples to complex-ecosystem-forest-peoples will not be easy, not least because it will be difficult to convince the isotropic-plains-peoples that they really *want* to become forest-peoples. Large-scale reforestation will challenge existing governance systems and conventional wisdoms, but that is at least partly the point.

What does the research on the matter suggest? I am not aware of any attempts at cost-benefit analyses of industrial agriculture on the scale of the Stern Review (2007). Such an analysis would attempt to compare the current command-and-control type of industrial agriculture with a theoretical shift toward a more local or community-based agroforestry model. Variables to be considered would have to include the productive capacity of a livable climate and healthy soils, which are currently being systematically exhausted. As argued in earlier chapters, such a comparison based solely on the quantity of food produced would be entirely inadequate and misleading. Such a framing might simply represent a leap outside our cultural norms, but knowledge of the consequences of current land-use patterns is certainly available. We might begin to assemble it by looking at the historical relationship between the traumatized mind that gave rise to Western civilization and the forests. According to Harrison (1992), "the first antagonist of Gilgamesh is the forest" (p. 14), the historical roots of the Western antipathy toward forests run very deep indeed.

This juxtaposition of forest as the symbol of the uncontrolled and fearful with the city as the symbol of human control and mastery is the very origin of unsustainability. Schama (1995) confirms and extends the work of Harrison:

> By the time William the Conqueror arrived on the Sussex coast, no more than 15 percent of English territory would have been wooded … By the time of Anglo-Saxon kings, then, the essential familiar pattern of the English countryside—broad tracts of cultivated field and pasture punctuated with copses and limited stands of trees—had already been established. There were still substantial areas of woodland in counties like Middlesex and Warwickshire, but from the fifth to the eleventh century they were steadily shrinking as clay soils were taken for farming. By the time of

> the Doomsday Book in 1086, areas whose very names signified woodland, like the Kentish Weald, had been converted into pasture, orchard, and arable. (p. 144)

This was the material context for the psychogenetic origins of the marketing mind and the capitalist economic system just as the stone hand axe was the material context for the evolution of the modern human-cognitive system.

As usual, wherever the Western-Cartesian mind is let loose on the world, it falls apart leaving peoples with few options other than a confused and disorganized retreat from their war on nature, usually followed by a re-grouping for a more concerted assault. Nevertheless, Louv (1998) describes plans in England and Scotland to plant "1.3 million trees ... to cool the urban heat of our island, provide shade, reduce stress, and increase activity" (pp. 86–87). These efforts to establish forests for the mental and physical health of peoples are occurring now in many countries around the world. This should come as no surprise. Respite from war often provides psychological benefits. Côté-Lussier et al. (2014) found, for example, that "targeting residential neighborhood features such as greenery and disorder could improve youth's felt safety at school, particularly for those experiencing chronic and later-childhood poverty." (p. 1195). This Quebec study found that the amount of greenery in the neighborhood of children experiencing poverty moderated feelings of insecurity and danger. In a world of increasing insecurity and danger, the expected psychological benefits of massive reforestation campaigns should provide all the justification that policy makers and regional planners need for a radical rethinking of current practices. If we further consider how deforestation has helped to support and maintain the marketing mind, then I think the weight of evidence in favor of reforestation becomes simply overwhelming.

But, in case the reader remains unconvinced, consider how a massive reforestation campaign would impact global food-production systems. Raising food from forests would entail many radical differences from the current industrial model, not least of which would be the shift in diet. Many trees provide edible nuts, fruit, or berries and some provide medicinal products or other products such as maple syrup. There are also many shade-tolerant, understory, fruit-bearing shrubs and woody plants such as currants, gooseberries, blueberries, elderberries, raspberries, blackberries, and many others. Vegetables eaten for their leaves or roots tend to be fairly shade tolerant also: sorrel, lettuces, mustard greens, spinach, chards, parsnips, carrots, beets, garlic, potatoes, and many others. Some spices and medicines such as ginseng, ginger, or tumeric do exceptionally well in forest conditions. Many edible mushrooms can also be grown in forests. Other sun-loving vegetables can be grown on rooftop gardens. In tropical regions, under climate-change scenarios where nighttime low temperatures may begin to stay above 30 degrees Celsius for many consecutive days, agroforestry will become increasingly important as most conventional commercially grown food crops cannot be economically sustained under such conditions.

By "reforestation," obviously, I am not talking about monocultures of trees in straight lines. A complex forest ecosystem provides benefits across and throughout extended and distributed ecologies of mind. The forest can be the best partner for healthy human development. If that relationship can be fostered and developed, it would not be unreasonable to expect that at some point a sustainable relationship could be established with animals to provide meat from rabbits, grouse, turkeys, and deer. Such a relationship would not need to include guns, and would constitute a vast improvement on our current relations with the animals that provide us with dietary protein while remaining essentially invisible. It would not replicate the pressure-cooker style of disease incubator that the current relationship entails (Hinchliffe, Bingham, Allen, & Carter, 2017). These benefits could also be expected to extend into fish habitats, eventually providing additional sources of dietary protein. Industrial processes also need to evolve within the materially engaged ecology of mind so these food sources become safe to eat and not contaminated with industrial toxins.

It is important to also consider how a society's cultural tools might be expected to evolve under a regime of large-scale reforestation. The plow would drift toward obsolescence. Economies could be more effectively localized. Economic knowledge would become more evidence based and less mythological. Our experience of the world would be drawn toward the local and the experiential as well. Our senses would be educated less by the materialized ideology of anxiety and control and more by living environments that could be more easily conceptualized as partners, if not blood relations. The list of potential advantages is long. In regions such as western North America or central China, where melting glaciers are expected to cause severe water shortages in coming decades, massive reforestation efforts might represent an option to reduce piping water over long distances and replenish shrinking ground water reserves to extend the habitability of places like Los Angeles, Calgary, or Beijing into the next century. Forests would absorb greenhouse gases and become a mitigating agent of global climate change if the planet were even reforested to roughly approximate the forests extant in 1800. The psychological benefits of trees have only recently begun to be studied rigorously, but it has become clear that trees in the landscape provide enormous psychological and mental-health benefits.

According to Lopes (2017), "repair-knowledges of many kinds need to be recovered, as do ways of teaching and learning them, and nurturing a desire for them. More usually these dispositions are exercised in the practice of clearing, cleaning or maintenance" (p. 188). Repair-knowledges represent a fundamental ontological re-orienting of our relations with the world and apply as much to reforesting the world as to de-plasticizing it. Repair provides an alternative way of being in the world, a different way of looking at a mowed lawn or a lake whose sufferings become apparent to us through our interactions with it. This is not the kind of knowledge that can be "learned" in a school without an ontological shift that includes a new answer to the question, "What is a student?" Lopes (p. 189) goes on to explain, "As a practice of sustainment, repair brings an 'aura' of worldliness that reminds us: 'I am not a human being in general but a particular human being in a particular historical community.'"

Conclusion

These examples of perceptual, cognitive, behavioral, and environmental reorientations that could be expected to accompany any large-scale shift away from the Western-Cartesian delusion toward a relational-ontology model such as MET could be multiplied exponentially. Everything would change—first the zig, then the zag, and then back and forth between the two until the old Cartesian world is no longer recognizable. The increasing porosity of ourselves and our worlds, the blurring of boundaries and the coming into focus of interconnections, opens the world to new ways of perceiving and understanding. As we come to recognize our minds extending into the world and the world extending into our minds, the old ways of thinking about ethics and about our taken-for-granted ways of relating to the world and to each other, in short *everything*, is transformed. But the change cannot simply be contained within the Cartesian confines of the ghost in a machine. It is fundamentally *not* about a different *knowledge* alone. If that relational ontology is not materialized in the world, then the world will continue projecting our own Cartesian insanity back to us and our escape from unsustainability will be foreclosed, possibly forever. There is no salvation for ourselves without first healing the waters, the lands, the air, and all the tools and other things we make and engage our brain/bodies with.

It was not an accident that much of this book is composed of personal experiences. It is critical to share such experiences, to become much more self-reflective, and to wonder at the horrifying incomprehensibility of our society's acquired incapacity to rise to the challenge of sustainability. Otherwise, we risk settling for the kinds of Band-Aid solutions to symptoms that leave broader and deeper underlying pathologies left to face, as in such examples as the Inquisition, genocides of indigenous peoples, or the abolition of slavery. All these abominations were preceded and followed by other atrocities reflecting materializations of the more systemic underlying cycles of trauma and social distress. Their ongoing impacts, in other words, continue to be felt.

There is much new research to support the shift to a post-Cartesian ecology of mind, and it is important to make a concerted push to see the transition through to the other side. Barrett (2011) is representative of a growing body of literature to argue convincingly that

> we augment, enhance, and support our brain-based internal learning to an extraordinary degree, hauling ourselves up by our bootstraps to achieve feats that no one individual could achieve alone. The things that make us smart in our peculiar way now occur outside our heads, as much as in them. (p. 222)

Unfortunately, in the globalizing Western-Cartesian paradigm, smart seems to mean suicidal and sadistic. As Barrett goes on to clarify, "The world-as-it-really-is need not, and probably does not, conform to the categories we impose on it, especially given the manner in which language and culture have allowed us to transform our environment to such an unprecedented degree" (p. 223). And this brings us back to the very beginning in Chapter 1, to how we might find the "right" answers to ontological questions such as, What is a human? What is a tool? or What is the world?

The choice before us now as a society is between the continued destruction of ourselves and the world or the establishment of a co-constitutive partnership with the world. It is not a choice we alone can make. The world also has its own ecology of mind and any partnership would need to be on terms not dictated by us. The "things that make us smart in our peculiar way" (Barrett, 2011, p. 222) are not directly within our control at any given moment in time but are, increasingly, the materialized insanities of a historical inheritance that pre-exists us. We should perhaps not expect to be given mercy from a world that we have shown very little mercy to. There are increasingly obvious signs we may have over-stayed our welcome, that the parent-child relationship has become frayed and may soon be severed. When the Golden Rule can no longer be deciphered from the rocks, trees, fishes, or birds; when conscious thoughts can no longer be entertained unless they lead to obvious economic advantage; when every breath we take incorporates plastic nanoparticles into our lung tissues; or when children become like raw materials for a refinement process designed to maximize their market value, then it might just be time to admit that "our peculiar way" has no rightful place in the world anymore. It may not yet be too late, but given the nature of the Western-Cartesian historical response to self-induced catastrophe, the odds are not working in our favor. The insanity of our age is not unprecedented in its degree, only in its scale. What we need is an epidemic of radical sanity and eco-mental health. For that to occur, I would suggest that we start by looking to the traces.

References

Barrett, L. (2011). *Beyond the brain: How body and environment shape animal and human minds*. Princeton, NJ: Princeton University Press.

Barry, J. (2012). *The politics of actually existing unsustainability: Human flourishing in climate-changed, carbon-constrained world*. Oxford, UK: Oxford University Press.

Boulding, K. E. (1978). *Ecodynamics: A new theory of societal evolution*. Beverley Hills, CA: SAGE.

Bruner, J. (1986). *Actual minds, possible worlds*. Cambridge, MA: Harvard University Press.

Clark, M. C. (2002). *In search of human nature*. London, UK and New York, NY: Routledge.

Côté-Lussier, C., Barnett, T. A., Kestens, Y., Thanh Tu, Y., & Seguin, L. (2014). The role of the residential neighborhood in linking youths' family poverty trajectory to decreased feelings of safety at school. *Journal of Youth and Adolescence*, *44*, 1194–1207.

Fasenfest, D. (Ed.). (1993). *Community economic development: Policy development in the U.S. & U.K.* New York, NY: St. Martin's Press.

Gosden, C. (1994). *Social being and time*. Oxford, UK: Blackwell.

Gosden, C. (2005). What do objects want? *Journal of Archaeological Method and Theory*, *12*, 193–210.

Gowdy, J. M. (1994). *Coevolutionary economics: The economy, society, and the environment*. Boston, MA: Kluwer Academic.

Greer, J. L., & Gonzales, O. (2017). *Community economic development in the United States: The CDFI and America's distressed communities*. New York, NY: Palgrave Macmillan.

Haraway, D. (2016). *Staying with the trouble: Making kin in the Chthulucen*. Durham, NC: Duke University Press.

Harrison, R. P. (1992). *Forests: The shadow of civilization*. Chicago, IL: University of Chicago Press.

Hinchliffe, S., Bingham, N., Allen, J., & Carter, S. (2017). *Pathological lives: Disease, space and biopolitics*. Malden, MA: Wiley.

Hodgson, G. M. (2001). *How economics forgot history: The problem of historical specificity in social science*. London, UK: Routledge.

Keating, G. C. (2018). Is cost-benefit analysis the only game in town? *Southern California Law Review, 91*(2), 195–262.

Latouche, S. (2009). *Farewell to growth*. Cambridge, UK: Polity.

Lopes, A. M. (2017). Ontological design as an ecological practice. *Arena, 47/48*, 172–191.

Louv, R. (2011). *The nature principle: Reconnecting with life in a virtual age*. Chapel Hill, NC: Algonquin Books. (Original work published in 1998).

Mendell, M. (1983). *Social, political and cultural determinants of economic activity: Comparative perspectives*. (Unpublished doctoral thesis). McGill University, Montreal, QC.

Nelson, M. K. (2017). Education for the eighth fire: Indigeneity and Native ways of learning. In E. Assadourian & L. Mastny (Eds.), *EarthEd: Rethinking education on a changing planet* (pp. 49–60). Washington, DC: Worldwatch Institute.

Polanyi, K. (1968). Primitive, archaic, and modern economies. Garden City, NY: Anchor Books.

Polanyi, K. (1977). The livelihood of man. New York, NY: Academic Press.

Schama, S. (1995). *Landscape and memory*. New York, NY: Vintage Books.

Schumacher, E. F. (1974). *Small is beautiful*. London, UK: Abacus.

Shragge, E. (2003). *Activism for social change: Lessons for community and local organizing*. Peterborough, ON: Broadview Press.

Shragge, E., & Toye, M. (Eds.). (2006). *Community economic development: Building for social change*. Sydney, NS: Cape Breton University Press.

Stern, N. (2007). *The economics of climate change: The Stern review*. Cambridge, UK and New York, NY: Cambridge University Press.

Swentzell, R. (1997). Conflicting landscape values: The Santa Clara Pueblo and Day School. In P. Groth & T. W. Bressi (Eds.), *Understanding ordinary landscapes* (pp. 56–66). New Haven, CT: Yale University Press.

INDEX

A

B

D

E

F

G

H

I

J

K

L

M

N

O

P

Q

R

S

T

U

V

W

Y

Z

www.ingramcontent.com/pod-product-compliance
Ingram Content Group UK Ltd.
Pitfield, Milton Keynes, MK11 3LW, UK
UKHW021835100926
8556IPUK00004B/351

9 781433 176661